Sustainability Outreach in Developing Countries

Mir Sayed Shah Danish · Tomonobu Senjyu ·
Najib Rahman Sabory
Editors

Sustainability Outreach in Developing Countries

Springer

Editors
Mir Sayed Shah Danish
Strategic Research Projects Center
University of the Ryukyus
Nishihara, Okinawa, Japan

Tomonobu Senjyu
Faculty of Engineering
University of the Ryukyus
Nishihara, Okinawa, Japan

Najib Rahman Sabory
Department of Electrical and Electronics
University of the Ryukyus
Nishihara, Okinawa, Japan

ISBN 978-981-15-7178-7 ISBN 978-981-15-7179-4 (eBook)
https://doi.org/10.1007/978-981-15-7179-4

This Springer imprint is published by the registered company Springer Nature Singapore Pte Ltd.
The registered company address is: 152 Beach Road, #21-01/04 Gateway East, Singapore 189721, Singapore

Preface

International Conferences Series on:
Sustainability Outreach in Developing Countries (SODC 2020)
May 30–31, 2020, Okinawa, Japan

Sustainable Development Goals (SDGs) adaption models and scenarios are interlinkage efforts, which are connected in some way to transform the world and secure societies' resiliency. Conceptualization of optimal models and scenarios requires impactful global endeavors to outreach viable sustainability. Deployment of interdisciplinary themes based on decent research, modern technologies, and innovative approves are necessary actions to put forward. Assessing alternative paths to the SDGs by implying potential synergies of incorporated multidimensional themes in terms of validating of finding and responsiveness is known exigence.

This book constitutes proceedings of the International Conferences Series on: Sustainability Outreach in Developing Countries (SODC 2020), held at University of the Ryukyus, Okinawa, Japan, from May 30 to 31, 2020. All submissions to this proceedings are subjected to double-blind peer review.

Twelve related full papers of the SODC 2020 processing provide an exhaustive overview of sustainability in the context of recent challenges, created windows of opportunities, expert experiences, case studies, lessons learned, and diversity of sustainability pillars (environmental sustainability, technical and technological sustainability, social sustainability, institutional sustainability, and economical sustainability) deployment in developing and developed nations.

The present book emanates recent researches based on an exhaustive topic of the day "Sustainable Development Goals (SDGs)", deals with transboundary researches, experiences, case studies, and lessons learned to cover conceptual and empirical research contributions within the scope of SDGs. Also, this book establishes interdisciplinary coverage of sustainability and argues its pillars (environmental, technical and technological, social, institutional, and economic disciplines), aligned with the 17 goals and 169 targets of SDGs for long-run sustainability.

The Sustainability Outreach in Developing Countries (SODC) conferences series is organized under the guidance of the Research and Education Promotion Association (REPA) a non-profit entity registered in Japan—Reg. No. 3600 05 006134 (www.repa.jp).

Nishihara, Japan
May 2020

Mir Sayed Shah Danish
Tomonobu Senjyu
Najib Rahman Sabory

Acknowledgements

The chairmen of International Conferences Series on: Sustainability Outreach in Developing Countries (SODC 2020) expresses its heartiest congratulation for successfully holding the conference. Thank you to our stakeholders at any level (volunteers, researchers, presenters, participants, reviewers, editors, facilitators, publisher-Springer, and sponsors) who have been part of this achievement. A special appreciation and thanks are extended to:

Keynote Speakers

Tomonobu Senjyu
Ir. Dr. Sharifah Rafidah Wan Alwi
Toshihisa Funabashi
Fathelalem Ali
Hamidullah Farooqi
Abdul Hamid Helmandi
Abdul Twab Balakarzai
Zalmai Zaheb
Mohammad Aref Naimzad
Mohammad Wasim Iqbal
Mohammad Naim Azimi
Abdul Qayoom Karim
Hameedullah Zaheb
Orzala Ashraf Nemat
Mohammad Ajmal Shinwari
Najib Rahman Sabory
Abdul Ehsan Mohmand
Sayed Hashmat Sadat
Aziz Azimi

Advisory Committee

Tomonobu Senjyu
Naomitsu Urasaki
Toshihisa Funabashi
Atsushi Yona

Volunteer Committee

M. Abdullah Omar
M. Hares Zaheer
Nasim Mahdi Shakery
Hanzala Quraishi
Abdul Sami Rahmani
Zabihullah
Sayed Ghafar Hashemi
Khalil Yusofi
Weqar Amin
Sediqullah
Sultan Hossain
Mustafa Arian
Heela Anvari
Firoza Aryan
Sumaia
Shahab Ahmad
Abdul Rahman Noori
Hasibullah Mohammadi
Yasamin Ghayasi
Lila Rawi
M. Mustafa Arefi
Sweeta Faizi
Suhib Azimi
Narwan Hofiany
Muzhda
Ahmad Milad
Shamim Bahadury
Zakria Afshar
Abdul Khaliq
Esmatullah
Sarem Shah
Ahmad Iqbal Habibi
Abdul Matin Waziry

Lida Rajabi
Khalid Abbasi
Sameer Sarwary
M. Omar Rasouli
Frahnaz Nazari
Nargis Samimzada
Masooma Saberi
Fayaz Shams
Ahmad Fariullah Omid
M. Hassan Sarvari
Khwaja Yahya Sediqi
Fayaz Rukhaye
Mohammad Omar Rasouli
Nazar Gul
Muzhgan Karimzada
Sonam Qaumi

Sponsors and Facilitators

Research and Education Promotion Association (REPA)
Rahkaar Research and Education Organization
University of the Ryukyus
Kabul University
IEEE-SEIES
Afghanistan Research and Evaluation Unit (AREU)
Amiri Medical Complex
TEchnologists, Inc. (Ti)
Rahkaar ACM Chapter
Dynamic Vision (DV)
Energy and Mining Research and Services Center (EMRSC)
Rahkaar IET on Campus
Water-Center Afghanistan
Also, heartfelt thanks for the editorial team of REPA: Michell Ann (Secretary), Emil Chuck, Sarah Ahmed, and Kristina Barnes for their support in administrative, management, copyediting, typesetting, and proofreading of submissions.

Nishihara, Japan
May 2020

Mir Sayed Shah Danish
Tomonobu Senjyu

Contents

Editors and Contributors

About the Editors

Assistant Professor Mir Sayed Shah Danish (Dr. Eng., MBA, CEng., SMIEEE, MIET) has been an engineering and technology expert and an academician for several years. He demonstrates a simple, in-depth style of narration of concepts, turning concepts into measurable endeavors and exploring interdisciplinary coverage in a systematic manner. Apart from being a scientific scholar, he brings together multidisciplinary skills and expertise (energy, environment, business, and management) providing integrated solutions. He is the author of several academic and technical textbooks, guidebooks, training manuals, and other books in English and Dari (Persian) languages. These publications have enabled him to link industry with academia, and he has achieved recognition with several awards and expressions of appreciation.

Since 2004, he has been involved in multidisciplinary engineering and technology by leading several projects in those fields while continuing as an active scholar and educator. He is an assistant professor at the University of the Ryukyus, Japan; founder and chair of the IEEE-Sustainable Energy and Intelligent Engineering Society (SEIES-PES & FRID joint chapter, Fukuoka Chapter); founder and facilitator of the Rahkaar IET On Campus Society; founder of the Rahkaar Research and Education Organization; and founder and president of the Research and Education Promotion Association (REPA).

He has worked with national and international organizations and companies as an urban electric power planner, team leader, technical advisor, department head, educational manager, and director. He is a chartered engineer, UK (CEng.), senior member of IEEE, member of IET (MIET), and holds membership in many other academic societies. He received his bachelor's degree in electrical and electronic engineering (B.Sc.) in 2009 from Kabul University, Afghanistan; two master's degrees, one in energy and electrical engineering (M.Sc.) in 2015 from the

University of the Ryukyus, Japan; the other in business administration (MBA) in 2016 from the National Institute of Business Management, India; and a doctorate in sustainable energy (Ph.D.) in 2018 from the University of the Ryukyus. He chaired and has been a committee member of several conferences and symposiums and has more than 50 publications. His main research interest is sustainable energy (policy, economics, market, environment, and management), smart cities and housing, storage systems, voltage stability, and related areas.

Professor Tomonobu Senjyu was born in Saga Prefecture, Japan, in 1963. He received his B.S. and M.S. degrees in electrical engineering from the University of the Ryukyus, Nishihara, Japan, in 1986 and 1988, respectively; and his Ph.D. degree in electrical engineering from Nagoya University, Nagoya, Japan, in 1994. He is currently a full professor with the Department of Electrical and Electronics Engineering, the University of the Ryukyus. He is an active scholar who makes not only theoretical and practical contributions to the industry but, as well, helps to produce valuable human resources in his graduates. Most of his students are from developing countries, pursuing their studies under Japanese government scholarships. He has supervised tens of Ph.D. dissertations since 2004. His former students are now actively involved in industrial and academic fields throughout the world.

He leads the Power Energy System Control Laboratory (PESC), which employs recent technology and produces novel research outcomes. His laboratory publishes several peer-reviewed journals each year. He has made salient contributions to many international journals and has more than 500 peer-reviewed publications in high-ranking academic databases to his credit. His research interests are in renewable energy, power system optimization and operation, power electronics, and advanced control of electrical devices.

Associate Professor Najib Rahman Sabory graduated from electrical and electronics department of the engineering school at Kabul University in 2001. Since then, he has been teaching in this department. From 2009 to 2011, he served as the deputy dean for the engineering school at Kabul University. After a Fulbright scholarship was awarded to him in 2011, he completed a master's degree in sustainable energy and a graduate certificate in project management from A. James Clark School of Engineering at the University of Maryland College Park. He has also earned an MBA in 2014 from World Wide Science Academy in Malaysia. He was awarded a United Nations Institute for Training and Research (UNITAR) fellowship in 2007 for a period of seven months. He learned many advanced concepts in project management, leadership, and change management. He later remained in this program for 2008 and 2009 cycles as coach and resource person. He has participated in a series of workshops and symposiums about leadership and entrepreneurship back in USA. In 2011, he led an Inter-ministerial Commission for Energy (ICE) committee for preparing capacity building strategy for energy sector of Afghanistan.

Currently, he serves as an associate professor and head of the energy department of the engineering school at Kabul University. He is also the deputy chairman of

Afghanistan Renewable Energy Union (AREU) since early 2017. Moreover, he is the rotating chairman of Afghanistan Awareness and Analysis (A3), a Kabul-based Think-Tank.

Contributors

Milad Ahmad Abdullah Kabul University, Kabul, Afghanistan

Mohammad Hamid Ahadi Research and Education Promotion Association (REPA), Okinawa, Japan

Mikaeel Ahmadi University of the Ryukyus, Okinawa, Japan

Sayed Hujjatullah Ahmadi Kabul University, Kabul, Afghanistan

Mujtaba Alemi Kabul University, Kabul, Afghanistan

Ahmad Zia Amini Kabul University, Kabul, Afghanistan

Ahmad Rasa Arsallan Kabul University, Kabul, Afghanistan

Mohammad Qasem Azad Kabul University, Kabul, Afghanistan

Ujjal Chattaraj NIT Rourkela, Rourkela, Odisha, India

Mir Sayed Shah Danish University of the Ryukyus, Okinawa, Japan

Sayed Mir Shah Danish University of the Ryukyus, Okinawa, Japan

Abdolhamid Ebrahimi Kabul University, Kabul, Afghanistan

Nabila Fazli Kabul University, Kabul, Afghanistan

Habibullah Fedayee University of the Ryukyus, Okinawa, Japan

Kh Jamilurahman Habibizada Kabul University, Kabul, Afghanistan

Mozhdah Hafizyar Kabul University, Kabul, Afghanistan

Zakia Husssainy Kabul University, Kabul, Afghanistan

Abdul Matin Ibrahimi University of the Ryukyus, Okinawa, Japan

Mohammad Hafiz Kohnaward Kabul University, Kabul, Afghanistan

Himayatullah Majidi Kabul University, Kabul, Afghanistan

Hamid Maliki Kabul University, Kabul, Afghanistan

Ghulam Farooq Mansoor Health, Nutrition and Social Science Research, Freelance Researcher, Kabul, Afghanistan

Homaira Mansoor Kabul University, Kabul, Afghanistan

Fahim Momand Kabul University, Kabul, Afghanistan

Ahmad Ajmal Oryakheill Kabul University, Kabul, Afghanistan

Yasser Qudir Kabul University, Kabul, Afghanistan

Naimatullah Shafaq Rahmatyar Baghlan University, Pol-e-Khomri, Afghanistan

Bashir Ahmad Raqi Kabul University, Kabul, Afghanistan

Nazifa Rasoli Kabul University, Kabul, Afghanistan

Najib Rahman Sabory Kabul University, Kabul, Afghanistan

Tomonobu Senjyu University of the Ryukyus, Okinawa, Japan

Shawkatullah Shams Kabul University, Kabul, Afghanistan

Hameed Shirzad Kabul University, Kabul, Afghanistan

Mohammad Zubair Stanikzai Research and Education Promotion Association (REPA), Okinawa, Japan

Zahra Sufizada Kabul University, Kabul, Afghanistan

Atsushi Yona University of the Ryukyus, Okinawa, Japan

Hasina Zadran Kabul University, Kabul, Afghanistan

Energy and Environment Efficiencies Towards Contributing to Global Sustainability

Mir Sayed Shah Danish, Najib Rahman Sabory, Mikaeel Ahmadi, Tomonobu Senjyu, Himayatullah Majidi, Milad Ahmad Abdullah, and Fahim Momand

1 Introduction

Sustainable development concept was proposed in 1980 by the World Conservation Strategy [1] and developed by the Our Common Future in 1987 [2] that covers social, economic, and ecological sustainability requirements. The concept of sustainability defined "meets the need of present generation without compromising the ability of future generations to meet their own needs" by the Earth Summit in 1992 [1]. The concept of sustainability typically quoted that how to use the energy resources in a way to be sufficient for now, and do not compromise the ability of future generations to meet their needs [3], which is analyzed into five pillars (Fig. 1).

The sustainability scale establishes a degree of sustainability. The more the scales of the indicator, the higher the accuracy. There are social indicators that consider the social effects of renewable energy sources (e.g., job creation, residents benefitted). Economic indicators are needed to assess the economic effects on the evaluation of renewable energy sources; these include costs, return analysis, and payback period. Reliable energy provision for socio-economic development comes to attention. Energy insufficiency and poverty are linked that affects negatively on lifestyle and expectancy [4]. Within renewable energy systems, there are capital costs, replacement costs, as well as operation and maintenance costs [5]. The general sustainability indicators must provide users the average numerical value of the system's sustainability for comparison and selection (i.e., greatest sustainability with the lowest cost). Users must be positioned to better determine if a system's benefits exceed its costs

M. S. S. Danish (✉) · M. Ahmadi · T. Senjyu
University of the Ryukyus, Okinawa 9030213, Japan
e-mail: mdanish@lab.u-ryukyu.ac.jp

M. S. S. Danish · N. R. Sabory · H. Majidi · M. A. Abdullah · F. Momand
Kabul University, Kabul 1006, Afghanistan

M. S. S. Danish et al. (eds.), *Sustainability Outreach in Developing Countries*,
https://doi.org/10.1007/978-981-15-7179-4_1

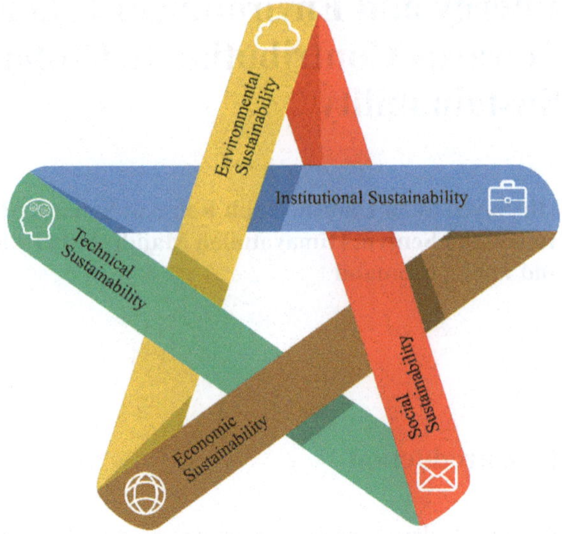

Fig. 1 Energy and environment within sustainability pillars

or vice versa. A numerical value is helpful to simplify the assessment and decision making.

However, an energy-efficient and sustainable infrastructure demonstrate with multidimensional parameters, according to [6]. Some essential criteria are considered as: sustainable energy production are accessibility, affordability, disparity, safety, use efficiency, supply and production efficiency, cost-effectiveness, and environmental impacts on air, water, and soil quality.

Sustainability within the energy sector requires an intensive study of multidimensional factors—some of which do not have a significant impact on climate change and global warming. Climate change variables must be identified to eliminate inadequacy and approximation of the analysis. There are significant issues that require in-depth consideration. In particular, global efforts are made to stem environmental and socio-economic devastation; nonetheless, root-cause analysis can omit these factors, introducing alternative approaches to overcome this complicated global challenge.

2 Sustainable Development Within Energy, Water, and Environment Constraints

The term of sustainable development has been used since decades that expanded its requirements from three pillars (environmental sustainability, social sustainability, and economic sustainability) to additional pillars as follows [7, 8]: From the perspective of energy and environment, execution of the sustainability concept encourages in terms of scenario-based analysis through the decision-making process to reach an optimum solution. Decision-making process for environmental-friendly energy system are listed in [9] as follows:

- Confirming the scenario score
- Ensuring state statements and goals
- Defining the indicators
- Evaluating indicators interconnection and impact
- Considering the hypothesis impact.

Weighing the indicators climate change and global warming are overwhelmingly influenced by human (i.e., anthropogenic) activities such as global industrialization and lifestyle modifications. People are over-exploiting natural resources while polluting the environment and destroying the ecosystem. This deadly combination has resulted in global energy demands that are rapidly escalating; the dependency on fossil fuels has led to enormous organic and non-organic waste production. Without conservation and preservation practices to mitigate these hazardous actions, the situation is worsening at an alarming rate. Energy efficiency and conservation concerned with reducing energy consumption in terms of resource and cost to boast climate change combat.

Anthropogenic activities like fossil fuel-consumption influences climate change. However, there are nonrenewable resources that are environmentally friendly and cost-effective, especially when considering alternatives such as biomass and waste-to-energy [10]. Another option is waste-derived fuels: solid fuels developed from various types of waste—both municipal and industrial; these fuels can be employed as a waste management option, providing diversions from landfills and their deadly methane gas emissions. This process mitigates greenhouse gases in the atmosphere, reducing climate change. Organic waste composting and recycling are additional alternatives in sustainable waste management, while biodiesel production is a significant renewable energy source.

All individuals throughout the world require energy and water to exist on this planet. Economic growth and human prosperity are strengthened by water and energy, which are closely interlinked and interdependent resources. These natural resources are considerably affected by the increased rate of urbanization, as societies put greater demands on finite supplies. A possible solution to address these problems is the use of wastewater treatment and reuse as a method to preserve water resources. Wastewater plants' main aim is to control water pollution and generate energy using digester gas ignition, which can potentially produce vast amounts of heat energy [10]. This treatment accelerates natural water purification processes, of which there are two stages. In the primary stage, solids settle and are then removed from wastewater; in the secondary stage, biological processes are utilized to further purify wastewater. Furthermore, these wastewater methods can be powered by alternative forms of electricity and heat (e.g., photovoltaic cells), creating a combined energy production system; this reduces the demand from grid energy, allowing the wastewater process to be contained entirely and energy-efficient. Environmental sustainability aims to ensure that human welfare has been improved. For this goal to be met, there are four environmental sustainability criteria (Fig. 2).

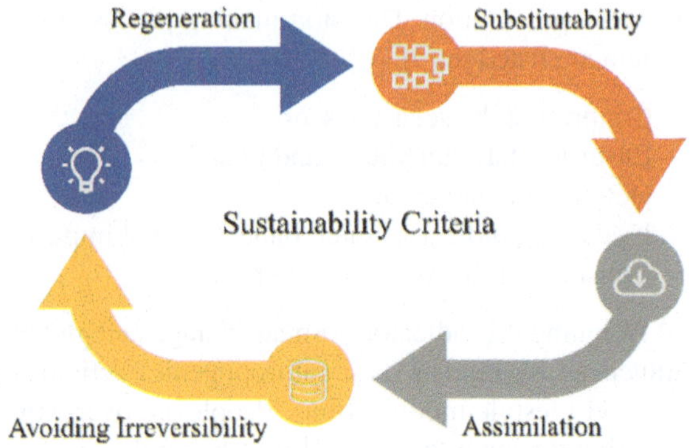

Fig. 2 Sustainability criteria [10]

In regeneration, renewable resources are used efficiently, not exceeding the long-term rates of natural regeneration [10]. In substitutability, nonrenewable resources are also used efficiently. Assimilation refers to limiting the release of environmental pollutants in relation to their waste assimilative capacity. Finally, irreversibility must be avoided [10]. The above standards are used to envision five interrelated objectives that advance environmental policies in the context of sustainable development. These include maintaining ecosystem integrity by pursuing viable options [11]:

- Efficiently managing of natural resources
- Decoupling environmental pressures from economic growth
- Enhancing the quality of life
- Improving global environmental interdependence through enhanced governance and cooperation
- Measuring progress using environmental indicators and indices, and
- Sustainable decision-making.

3 Measuring Environmental Sustainability Performance

For the development of environmental sustainability, the importance of ecosystem services can be utilized in which reinforces the value of non-monetary ecological qualities and functions. For a country to be able to provide transparent, objective methods of measuring and demonstrating energy and environmental sustainability, indices and indicators must be employed [3]. Assessments of environmentally-friendly sustainable energy incorporate both anthropogenic and natural activities, focusing on the long-term impacts and social justice issues for both current and future generations [12]. Energy, including renewable and geological storages, is an essential input to all forms of economic and social activities.

The environmental indices must be successfully implemented, considering essential factors such as the impacts of energy demand increases and economic activities.

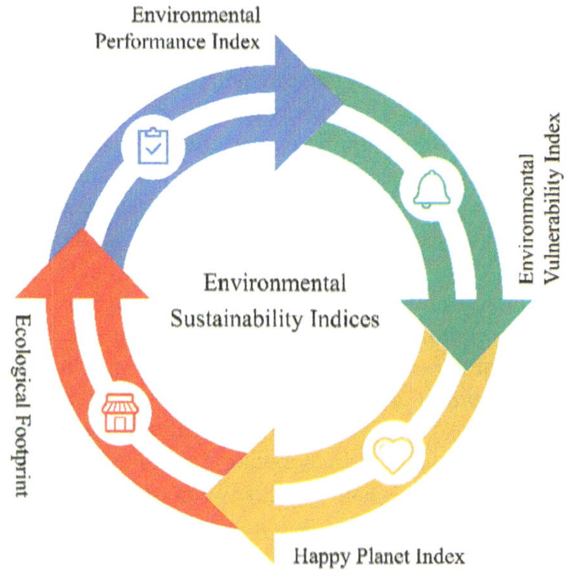

Fig. 3 Environmental sustainability indices

Other variables include the effects of resource productivity and environmental dilapidation on economic productivity. Finally, there are influences on society resulting from environmental improvements, such as enhanced human well-being—physical, psychological, emotional, and spiritual. Not only do improvements enhance the world in which people live, but they also add to the aesthetic value that is placed on them.

According to Olafsson et al. [12], four environmental indices identify relevant variables for evaluation before data is collected and analyzed, integrating standards against which performance can be compared (Fig. 3).

The Environmental Vulnerability Index assesses the exposure of the physical environment to 50 environmental indicators, including risks, susceptibility to damage, and outcomes. The Environmental Performance Index attempts to quantify and compare countries' environmental performance using various indicators, such as reducing environmental health stressors as well as protecting ecosystems and natural resources. The Ecological Footprint calculates the total amount of goods and services consumed by a country's inhabitants on a per capita basis. At the same time, the Happy Planet Index represents a ratio of human well-being to environmental impact. It assesses the ecological efficiency to deliver a certain level of biased human well-being.

4 Exergy for Environment, Ecology, and Sustainable Development

According to Bilgen and Sarıkaya [13], exergy is the result of substance interactions in the natural environment—the maximum useful work material can perform while achieving equilibrium with a heat source. This work can be obtained from reversible steady-state flow processes. Exergy is a qualitative measurement, as opposed to energy's quantitative one. It is beneficial for evaluating energy flows in residential, industrial, and agriculture practices, improving designs, and minimizing energy usage. Findings from exergy analysis often lead to the adoption of "green" energy technologies that improve energy efficiency. Referring to the connection between the second law of thermodynamics and environmental impact, when energy efficiency is increased, both energy losses and environmental effects are decreased [13].

Various sectors have different impacts on the environment and climate change due to these sectors' energy consumption and generated wastes. The residential sector consumes about 42% of the total energy in the world [14, 15] that mainly fed by vitalization loads [16]. The transport sector has a considerable impact on the environment. All its energy needs are met through the burning of fossil fuels, contributing to air pollution from both direct and indirect sources. To adequately address this issue, alternative means of transportation can be implemented—such as cycling, walking, and carpooling. For the economy to benefit from these transport adjustments, the relationship between exergy and environmental problems must be analyzed [13].

Through utilizing mathematical models, exergy assessment helps explain important ecosystem features, resolving numerous difficulties associated with energy efficiency. Exergy is the measure of the distance from chemical equilibrium; it is used in the process of translating ecological indicators to thermodynamics. When environment and ecology are integrated with exergy, they offer a fresh approach to improving environmental and ecological management—providing significant performance potential [13]. The use of exergy improves technologies as well as sustainable energy applications, resulting in enhanced energy systems.

Industrial ecology provides principles as a tool for change; these include industrial symbiosis, technological food webs, closed industrial ecosystems, and industrial metabolism. Industrial ecology is employed to systematically analyze the interactions between the environment and anthropogenic activities. It is applied in the industrial sector to optimize the total industrial material cycle—from raw materials to finished products with less environmental impact. It is an environmentally-friendly method for industrial waste disposal. Industrial sector energy-related challenges are reported [17, 18]:

- Produce, transmit, and utilize energy in an environmentally friendly manner.
- Reduce overall costs by enhancement of generation and operation efficiencies.
- Diminish overall costs by the smooth implementation of management and business practices.

Exergy is an effective method for providing optimal environmental conditions. Exergy is a central concept in achieving sustainable development; one exergy-based indicator can be used to indicate the environmental effects associated with resource depletion and emissions. Obtaining sustainable development requires considerable reduction on exergy losses [13]. Exploitation of the renewable energy sources such as solar, wind, geothermal, biomass, hydrogen, and hydraulic has significantly increased in recent years. In part due to their distinct environmental advantages;

renewables reduce the demand for nonrenewable energy sources such as fossil fuels, thus decreasing resultant emissions and pollution. In developing energy sustainability, researchers must be aware of three general dimensions: ecology, economy, and society. These three elements depend upon each other for existence. Exergy analysis facilitates communication among these separate yet interconnected studies. It is essential to understand how exergy intertwines with the environment, ecology, and sustainable development [19].

Among potential strategies for sustainability outreach, exploitation of renewable energy technologies is known as a long-term solution.

Through demand participation, energy storage, promotes the use of hydropower and high enthalpy geothermal energy to meet growing power demands and other elasticity measures, a more flexible, smarter energy system is required to integrate large shares of renewables [20].

According to Cucchiella et al. [21], some indicators were selected for evaluation of country sustainability performances as follows:

- Greenhouse gas emissions
- Government expenditure on environmental protection
- Total recycled and reused waste from electricity, electronic equipment, and vehicles
- Total recycled materials and solid waste
- Renewable energy shares in electricity, transport, heating, cooling, and percentage change of primary energy consumption in a specific period.

Performed assessments in the literature demonstrated that environmental indicators are more valuable than energy indicators. The total recycled and reused waste indicator had a significant effect on the sustainability goal. The role of renewable energy in the heating and transportation sectors is an important consideration; however, renewables within the electrical energy sector contribute more fossil fuel reduction. In national energy strategies, low carbon technologies and energy efficiency are playing vital roles. The decline of environmental pollution and the growth of virgin resource conservation are potentially encouraged by the recycling of e-waste. A country is considered energy sustainable if its pollutant emissions are very low, and its energy consumption has improved over time.

5 Energy Generation, Transformation, Transmission, and Distribution Efficiency

Innovation in the energy sector has increased its supply cost while reducing energy costs. New inventions have made it possible to integrate various energy sectors, increasing efficiency within fossil fuel systems. It has also enabled the increase in renewable energy's share within the overall energy mix. The recent environmental and legitimate pressures employ the need for change in the world's energy mix [22].

This change anticipated at the global level and will be achieved at the national and local levels.

To satisfy the energy sector demands more efficiently, they must be optimally combined [23]. Long-term energy storage is assisted by transforming energy into different forms; for example, electricity can be converted into heat and chemical form, supporting long-distance energy transportation. Energy storage systems can be used to reduce cost, improve system stability, improve generation expansion (by using distributed generations), give manageability options, and to allow renewable sources dispatchability [4].

Voltage instability in a power system can swiftly lead it to voltage collapse and entirely blackout. Therefore continuous monitoring and control of voltage stability are indispensable [24]. Some obvious measures to control a power system instability are reactive power compensation, network loadability improvement, network re-configuration, and optimally distributed generations [25]. So, integration of distributed generation (DG) e.g., wind, solar, geothermal generation elements with a grid, can contribute to the system stability and control fluctuation nature of loads [26].

Stability in power generation systems can be negatively affected by increased distributions of unreliable renewable energy sources; this is due to the significant reduction of inertia and consequential risks on frequency control. In cases of high distribution, there are proposed approaches, which guarantee framework stability based on certain strategies. Electricity demand and supply from renewable sources can be forecasted. New or existing power generation units can be organized, and increased demand elasticity through appropriate load management strategies. Further, additional electricity uses, including direct or indirect energy storage technologies, can be introduced [20].

Through the first approach, efficient forecasts on the supply or demand side can be performed, addressing network stability issues in scenarios with high shares of renewable electricity; this minimizes the reserve capacity and associated cost. The probabilistic approach provides more quantitative information on the uncertainty associated with power generation, offering useful feedback for making decisions as compared to point forecasts [27].

For energy to be transported properly, high-quality infrastructure is required, since it is transported for long distances in different forms (e.g., chemical, electric, and thermal). Energy is primarily extracted, transported, and used in the process of conversion, which produces both electricity and heat. These are the two main uses of energy [23]. Technology has advanced within the energy sector, as renewable energy technologies gain large shares with proposals for new storage systems; these increase the system's flexibility [23].

A framework where various energy sectors interact with each other at different levels is represented by smart energy, or multi-energy, systems in the form of smart and microgrids. The microgrid term applies for a small scale energy/power system generation unit, involves multifarious typology and networks (interconnected, radial, and hybrid) [6]. A microgrid consists of generation, load, storage facilities, monitoring, control, and automation systems. The primary goal of energy infrastructures is to produce, transform, transport, and distribute energy. For studies in this field, approaches are developed for preferred designs and functions of integrated energy systems, guaranteeing accessible, reliable infrastructures.

Infrastructures are crucial for a global understanding of each separate system, especially its potential as an overall view of energy. The mutual influences of systems must first be realized before a mutually beneficial design—in terms of economic and thermodynamic competence and committed operations—can be created. Since complex components' behavior characterizes the energy infrastructure, it exhaustively investigated in the literature.

The main challenge within the current electric power grid is the increasing uncertainty rate, as the transition to new renewable sources of energy commences. With these new resources, the power grid system is forced to function as never before, meeting new standards and procedures. It operates closer to established safety boundaries, resulting in greater risks of synchrony loss and voltage collapse.

Recently, due to cooperative energy markets, power systems are impelled to operate with high efficiency (close to their collapse point of stability) that causes sensibility and risk of a system or part of a system blackout [28]. This new critical urgency has led to emerging development and implementation opportunities. It also, includes updated optimization and control schemes, which account for the underlying uncertainty, steering the system closer to its operational limits.

Moving through multi-energy structure asks an integrated concept of energy infrastructures. This integrated system offers more freedom and versatility, minimizing the overall consumption of primary energy resources by increasing the system's performance, enabling it to handle more renewable sources while better managing unforeseen events.

These innovative operations technologies, at both local and system levels, put much-needed pressure on the system. Authoritarian frameworks and new markets, not to mention unused resources, are employed to reduce the increased rate of stress. Pressure facilitates change. Joint operations of energy system grids may effectively manage these emergencies and increase optimum operations organization. This collaboration can lead to successful renewable energy systems that are optimally managed and executed.

6 Development of Sustainability Indicators and Renewable Energy

Sustainability indicators measure it reliably for renewable energy sources. Their main objective is to provide a broad and highly scalable, information-driven structure for sustainability measurement. Quantitative sustainability indicators are easily understood. For a long time, the attention of policymakers has been attracted by sustainable energy decision making, using multi-criteria decision analysis to provide an elimination method [29]. Identifying who is budgeting for a renewable energy system's sustainability and who will make necessary decisions [5]. When selecting the indicators, they should be based on practical terms, aligned with the sustainable development criteria within the following requirements (Fig. 4).

Before formulation and localization of sustainability indicators, understanding of these indicators' development and performance assessment processes is advisable. Among many approaches, suitability and pertinency of proposed approaches aligned with sustainability and efficiency criteria are essential. Assessing sustainability performance requires a transdisciplinary effort to expose origination,, benchmarks, parameters, impacts, performances, and, more importantly, a categorized data analysis. According to Danish [30], efficiency and sustainability indicators come under the fourth layers of data analysis (Fig. 5).

Offering a fitting proposal for environmental-friendly energy provision with optimum efficiency, viable strategies, policies, and process development requires the five pillars of sustainability, which are known exigence to achieve the SDGs

Fig. 4 Renewable energy decision-making indicators' requirement within sustainable development criteria [5]

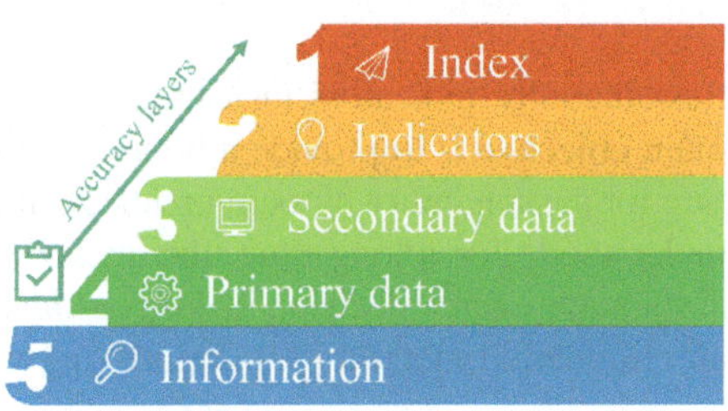

Fig. 5 Energy efficiency and sustainability indicators analysis theoretical analysis hierarchy

related goals [31]. Expert knowledge, tools, and suitable techniques can contribute to the process of a policy or a strategy development within scope and budget at constrained sustainability requirements [32]. This process can be purified by hiring strategic-interdisciplinary approaches of technical, technological, managerial, and policy development. Meanwhile, an immature policy/strategy can lead a nation to irrecoverable consequences that not only waste resources, also hinders the trends of development [33].

7 Conclusion

Increased concerns associated with the environmental impact of anthropogenic activities, compounded by energy security concerns and economics, are due to continuous population growth and rising living standards. From this review, it is apparent that the use of renewable energy sources will increase the energy sector's sustainability, reducing energy demands, and fossil fuel use. With decreased air pollution and greenhouse gases damaging the atmosphere, the harmful impacts of global warming and climate change can be lessened. This study outlines the main topics related to the energy and environment within sustainability constraints in the twenty-first century. This overview provides inside information in the context of sustainability and its indicators. Lessons learned and sustainability measures are considered to draw a big picture of the subject that can be counted as a reference for students, researchers, scholars, and field practitioners.

References

1. Moldan, B., Janoušková, S., Hák, T.: How to understand and measure environmental sustainability: indicators and targets. Ecol. Ind. **17**, 4–13 (2012). https://doi.org/10.1016/j.ecolind.2011.04.033
2. United Nations (UN), S.: Report of the World Commission on Environment and Development: Our common future. United Nations (UN) (1987)
3. Danish, M.S.S., Senjyu, T., Ibrahimi, A.M., Ahmadi, M., Howlader, A.M.: A managed framework for energy-efficient building. J. Bulid. Eng. **21**, 120–128 (2019). https://doi.org/10.1016/j.jobe.2018.10.013
4. Matin Ibrahimi, A., Masih Sediqi, M., Or Rashid Howlader, H., Sayed Shah Danish, M., Chakraborty, S., Senjyu, T.: Generation expansion planning considering renewable energy integration and optimal unit commitment: a case study of Afghanistan. AIMS Energy **7**, 441–464 (2019). https://doi.org/10.3934/energy.2019.4.441
5. Liu, G.: Development of a general sustainability indicator for renewable energy systems: a review. Renew. Sustain. Energy Rev. **31**, 611–621 (2014). https://doi.org/10.1016/j.rser.2013.12.038
6. Danish, M.S.S., Senjyu, T., Funabashia, T., Ahmadi, M., Ibrahimi, A.M., Ohta, R., Rashid Howlader, H.O., Zaheb, H., Sabory, N.R., Sediqi, M.M.: A sustainable microgrid: a sustainability and management-oriented approach. Energy Procedia **159**, 160–167 (2019). https://doi.org/10.1016/j.egypro.2018.12.045

7. Danish, M.S.S., Senjyu, T., Yaqobi, M.A., Nazari, Z., Matayoshi, H., Zaheb, H.: The role of ICT in corruption elimination: a holistic approach. In: 2018 IEEE 9th Annual Information Technology, Electronics and Mobile Communication Conference (IEMCON), pp. 859–864. IEEE, Vancouver, BC (2018). https://doi.org/10.1109/IEMCON.2018.8614890
8. Mohammadabadi, M.A., Ghoreshi, S.: Green architecture in clinical centres with an approach to Iranian sustainable vernacular architecture (Kashan City). Procedia Eng. **21**, 580–590 (2011). https://doi.org/10.1016/j.proeng.2011.11.2053
9. Danish, M.S.S., Matayoshi, H., Howlader, H.O.R., Chakraborty, S., Mandal, P., Senjyu, T.: Microgrid planning and design: resilience to sustainability. In: 2019 IEEE PES GTD Grand International Conference and Exposition Asia (GTD Asia), pp. 253–258. IEEE, Bangkok, Thailand (2019)
10. Urbaniec, K., Mikulčić, H., Rosen, M.A., Duić, N.: A holistic approach to sustainable development of energy, water and environment systems. J. Clean. Prod. **155**, 1–11 (2017). https://doi.org/10.1016/j.jclepro.2017.01.119
11. Danish, M.S.S., Sabory, N.R., Wali, M., Lotfy, M.E., Senjyu, T.: A sustainable building planning, modeling, and optimization within the smart city appraisal. In: International Journal on: Proceedings of Science and Technolgy. p. (in press). IEREK, Sepang, Malaysia (2019)
12. Olafsson, S., Cook, D., Davidsdottir, B., Johannsdottir, L.: Measuring countries' environmental sustainability performance—a review and case study of Iceland. Renew. Sustain. Energy Rev. **39**, 934–948 (2014). https://doi.org/10.1016/j.rser.2014.07.101
13. Bilgen, S., Sarıkaya, İ.: Exergy for environment, ecology and sustainable development. Renew. Sustain. Energy Rev. **51**, 1115–1131 (2015). https://doi.org/10.1016/j.rser.2015.07.015
14. Danish, M.S.S., Sabory, N.R., Ershad, A.M., Danish, S.M.S., Yona, A., Senjyu, T.: Sustainable architecture and urban planning trough exploitation of renewable energy. Int. J. Sustain. Green Energy 6, 1–7 (2017). https://doi.org/10.11648/j.ijrse.s.2017060301.11
15. Gopi, P., Reddy, I.P.: Modelling and optimization of renewable energy integration in buildings. In: International Conference on Sustainable Energy and Intelligent Systems (SEISCON 2011), pp. 116–120 (2011). https://doi.org/10.1049/cp.2011.0345
16. Ahmed, W., Fardan, H., Asif, M.: Integration of building energy modeling in the design process to improve sustainability standards in the residential sector—case study of the Eastern Province of Saudi Arabia. In: 2017 IEEE International Conference on Smart Energy Grid Engineering (SEGE), pp. 309–314 (2017). https://doi.org/10.1109/SEGE.2017.8052817
17. Akadiri, P.O., Chinyio, E.A., Olomolaiye, P.O.: Design of a sustainable building: a conceptual framework for implementing sustainability in the building sector. Buildings **2**, 126–152 (2012). https://doi.org/10.3390/buildings2020126
18. Dahunsi, F.M.: Conceptual framework for sustainable energy development in Africa. In: 2013 IEEE International Conference on Emerging and Sustainable Technologies for Power and ICT in a Developing Society (NIGERCON), pp. 238–241. IEEE, Owerri, Nigeria (2013). https://doi.org/10.1109/NIGERCON.2013.6715661
19. Rosen, M.A.: Engineering sustainability: a technical approach to sustainability. Sustainability **4**, 2270–2292 (2012). https://doi.org/10.3390/su4092270
20. Piacentino, A., Duic, N., Markovska, N., Mathiesen, B.V., Guzović, Z., Eveloy, V., Lund, H.: Sustainable and cost-efficient energy supply and utilisation through innovative concepts and technologies at regional, urban and single-user scales. Energy **182**, 254–268 (2019). https://doi.org/10.1016/j.energy.2019.06.015
21. Cucchiella, F., D'Adamo, I., Gastaldi, M., Koh, S.L., Rosa, P.: A comparison of environmental and energetic performance of European countries: a sustainability index. Renew. Sustain. Energy Rev. **78**, 401–413 (2017). https://doi.org/10.1016/j.rser.2017.04.077
22. Sagara, M., Shigenobu, R., Adewuyi, O.B., Yona, A., Senjyu, T., Danish, M.S.S., Funabashi, T.: Voltage stability improvement by demand response. In: TENCON 2017—2017 IEEE Region 10 Conference, pp. 2144–2149. IEEE, Penang, Malaysia (2017). https://doi.org/10.1109/TENCON.2017.8228215
23. Guelpa, E., Bischi, A., Verda, V., Chertkov, M., Lund, H.: Towards future infrastructures for sustainable multi-energy systems: a review. Energy **184**, 2–21 (2019). https://doi.org/10.1016/j.energy.2019.05.057

24. Danish, M.S.S., Yona, A., Senjyu, T.: A review of voltage stability assessment techniques with an improved voltage stability indicator. Int. J. Emerg. Electr. Power Syst. **16**, 107–115 (2015). https://doi.org/10.1515/ijeeps-2014-0167
25. Danish, M.S.S., Senjyu, T., Danish, S.M.S., Sabory, N.R., K, N., Mandal, P.: A recap of voltage stability indices in the past three decades. Energies **12**, 1544 (2019). https://doi.org/10.3390/en12081544
26. Furukakoi, M., Danish, M.S.S., Howlader, A.M., Senjyu, T.: Voltage stability improvement of transmission systems using a novel shunt capacitor control. Int. J. Emerg. Electr. Power Syst. **19**, 1–12 (2018). https://doi.org/10.1515/ijeeps-2017-0112
27. Bublitz, A., Keles, D., Zimmermann, F., Fraunholz, C., Fichtner, W.: A survey on electricity market design: insights from theory and real-world implementations of capacity remuneration mechanisms. Energy Econ. **80**, 1059–1078 (2019). https://doi.org/10.1016/j.eneco.2019.01.030
28. Furukakoi, M., Adewuyi, O.B., Danish, M.S.S., Howlader, A.M., Senjyu, T., Funabashi, T.: Critical boundary index (CBI) based on active and reactive power deviations. Int. J. Electr. Power Energy Syst. **100**, 50–57 (2018). https://doi.org/10.1016/j.ijepes.2018.02.010
29. Wang, J.J., Jing, Y.Y., Zhang, C.F., Zhao, J.H.: Review on multi-criteria decision analysis aid in sustainable energy decision-making. Renew. Sustain. Energy Rev. **13**, 2263–2278 (2009). https://doi.org/10.1016/j.rser.2009.06.021
30. Danish, M.S.S.: Green building efficiency and sustainability indicators. In: Senjyu, T. (ed.) Green Building Management and Smart Automation, pp. 128–145. IGI Global, Pennsylvania, United States (2020). https://doi.org/10.4018/978-1-5225-9754-4
31. Danish, M.S.S., Senjyu, T., Zaheb, H., Sabory, N.R., Ibrahimi, A.M., Matayoshi, H.: A novel transdisciplinary paradigm for municipal solid waste to energy. J. Clean. Prod. **233**, 880–892 (2019)
32. Danish, M.S.S., Zaheb, H., Sabory, N.R., Karimy, H., Faiq, A.B., Fedayi, H., Senjyu, T.: The road ahead for municipal solid waste management in the 21st century: a novel-standardized simulated paradigm. In: IOP Conference Series: Earth and Environmental Science, vol. 291, pp. 1–5 (2019). https://doi.org/10.1088/1755-1315/291/1/012009
33. Danish, M.S.S., Sabory, N.R., Danish, S.M.S., Ludin, G.A., Yona, A., Senjyu, T.: An open-door immature policy for rural electrification: a case study of Afghanistan. Int. J. Sustain. Green Energy **6**, 8–13 (2016). https://doi.org/10.11648/j.ijrse.s.2017060301.12

A Concise Overview of Energy Development Within Sustainability Requirements

Mir Sayed Shah Danish, Najib Rahman Sabory, Abdul Matin Ibrahimi, Tomonobu Senjyu, Mohammad Hamid Ahadi, and Mohammad Zubair Stanikzai

1 Introduction

Sustainability term is observed differently from various standpoints. From the energy and environment perspectives, sustainability can be described as the optimal use of energy resources sufficient for now, without compromising future generations needs [1]. It is proffered by many studies and echoed from different viewpoints, that achieving sustainable development will not be successful and viable unless full adoption of sustainability pillars are observed. Sustainability concept pillar are central and revolving around: environmental, ecological, social, cultural, economical, institutional, political, and technological sustainability [2]. Increased energy production and demand caused human beings with lifestyle changes and more productivity. However, energy generation and utilization have tuned societies for welcoming new crises, global warming and climate change. As a fact, energy is a globally conserved quantity, and it can neither be created nor destroyed as a universally constant amount. Energy security and balance come in the focus of attention for energy sustainability in the long-term.

In the country level, energy balance in supply and demands are composed of different factors such as supply and generation mixes, share of renewable and nonrenewable, self-reliance in supply, refining efficiency, energy transformation efficiency, per capita consumption of primary and final energy ration, energy intensity (based on the ratio of energy consumption to output of economic activities), energy

M. S. S. Danish (✉) · A. M. Ibrahimi · T. Senjyu
University of the Ryukyus, Okinawa 9030213, Japan
e-mail: mdanish@lab.u-ryukyu.ac.jp

M. S. S. Danish · N. R. Sabory
Kabul University, Kabul 1006, Afghanistan

M. H. Ahadi · M. Z. Stanikzai
Research and Education Promotion Association (REPA), Okinawa 9000015, Japan

M. S. S. Danish et al. (eds.), *Sustainability Outreach in Developing Countries*,
https://doi.org/10.1007/978-981-15-7179-4_2

equity, etc. [3]. Energy efficiency refers to saving energy for producing the same services our useful output [4]. In fact, sustainable development is a decision-making approach that balances today's human beings needs, respecting the demand of the future generations, with balancing social, economic, technical, institutional, and environmental requirements within a viable solution mechanism for long-term sustainability. Indeed, deployment of sustainability requirements can ensure socio-economic development along with improvement of lifestyle and elimination of natural energy resources (fossil fuel) from energy production and utilization lifecycle [5, 6].

Basically, energy defines as the capacity of a physical system to perform work. Or in different connotations, Work = Force × Displacement along the direction of force; or, work is the product of force and displacement through which the force acts. Where Force is the acting force in "Newtons" and *Displacement* is the displacement along the direction of force in (meters). Acting 1 N (N) force on a substance that displaces it a meter in the same direction performs an amount of work equivalent to one joule (J). It is also important to note that work, the capacity for doing work, and energy have the same units. A system may possess energy even when no work is being done. Since energy is measured by the total amount of work that the body can do, hence energy is expressed in the same unit of work as mentioned above. According to Tiwari and Mishra [7], forms of energy are shown in Table 1. For instance, converts chemical energy into electrical energy. The kinetic part of mechanical energy of a car converts into heat when the brakes are applied.

There is an famous law known as the "The Law of conservation of energy", states that the total amount of energy in a closed system remains constant. Energy may change from one form to another, but the total amount in any closed system does not change. This law is extremely important in order to understand a variety of phenomena related to energy and environment. The various forms of conversions of energy from one form to others are summarized in Table 2.

The 1970s energy crisis was almost forgotten by the 1980s that followed by environmental and global warming crises such as acid rain and radioactive waste as the main topic related to energy security [7]. Deployment of renewable energy technologies along with optimum operation of energy systems enables countries to assure energy security and mitigate climate change uncertainty [9]. From technical perspective, energy security concerned with reliable accessibility baked by stable operation at a cost-effective price and low greenhouse gas emissions [10].

The collection, management, and recycling of solid waste is a huge obstacle for many industries and residential. Research shows that recycling and composting solid waste reduce atmospheric carbon dioxide, primary greenhouse gas, and contributor to the climate change; therefore, these sustainable practices help decrease energy demand while conserving energy. Waste transportation and collection also emit greenhouse gases. To mitigate this hazard, a combination of home composting and fewer organic waste collection days can be very effective [11].

According to Bilgen and Sarıkaya [12], the most damaging environmental problems are acid rain, ozone depletion, and global warming. Acid rain occurs when sulfur dioxide and nitrogen oxides react in the atmosphere with water, oxygen, and other chemicals, forming acidic compounds; the reaction rate is accelerated by sunlight.

Table 1 Energy forms and their applications

Energy type	Explanation	Mathematic description	Abbreviation description
Kinetic energy (KE)	Possession of extra energy of motion: It defined as the acceleration of a body mass from rest to its current velocity [8]	$KE = \frac{mV}{2}^2(J)$	Where, m is the mass of the object in kg and v is its velocity in m/s
Potential energy (PE)	Possession of energy due to its elevation in a gravitational field	$PE = mgh(J)$	Where, m is the mass of object in kg, g is acceleration due to gravity in m/s and h is its height in m
Chemical energy	Possession of energy in molecules that can be arises due to atoms combination or separation (chemical interactions)		
Electrical energy	Defines as the capacity of moving electrons to evolve heat, electromagnetic radiation and magnetic fields		
Heat (thermal) energy	Possession of energy of a material due to the random motion of its particles		
Radiant energy	Radiant energy is the energy emitted by electrons as they change orbit and by atomic nuclei during fission and fusion; on striking matter, such energy appears ultimately as heat		
Nuclear (mass) energy	Possession of energy stored in the nucleus of an atom. According to Einstein, when the mass of some system is reduced by an amount Δm, as in nuclear reaction, then the amount of energy release	$E = \Delta mc^2$	Where, c is the velocity of light ($3 \times 10^8 m/s$)

Table 2 Energy conversion and types of utilization

	To mechanical	To electrical	To heat	To light	To mechanical
From light	Plant (photosynthesis)	Solar cell	Heat lamp radiant solar	Laser	Photoelectric and door opener
From chemical	Food plants	Battery fuel cell	Fire food	Candle phosphorescence	Rocket animal muscle
From electrical	Battery electrolysis electroplating	Transistor transformer	Toaster heat lamp spark plug	Fluorescent lamp and light-emitting diode	Electric motor relay
From heat	Gasification vaporization	Thermocouple	Heat pump and heat exchanger	Fire	Turbine gas engine and steam engine
From mechanical	Heat cell (crystallization)	Generator alternator	Friction brake	Flint spark	Flywheel pendulum and water wheel

Acid rain damages plant leaves by reduces soil nutritional levels [12]. Using alternative energy sources other than fossil fuel benefits the environment, and lessening acid rain's harmful effects. In ozone depletion, ozone-depleting substances (ODS) are released into the atmosphere through industrial practices. The most harmful ODS is chlorofluorocarbons. The ozone layer helps balance the earth's energy, absorbing dangerous solar ultraviolet radiation; when it is depleted, this radiation makes it through the earth's surface. Human health is adversely impacted. Finally, the third environmental issue is the global warming. This occurs in certain geographical areas where substantial temperature fluctuations occur, perpetuating the process of climate change. Three significant challenges associated with climate change that policies must address, are emissions reductions, technological innovations, and preparedness.

2 Energy and Human Needs

Increasing growth in pollution, lifestyle changes, negative global interactions, inadequate policies, self-determination of governments, and rapid consumption of natural resources are the rudimentary drivers towards climate change and global warming. Yet natural energy resources consumptions remain challenging efforts that couldn't pave the way for a global consensus that affected by the economic and military superpowers in the world. Recalling the energy crisis in the 1970s, followed by the 1980s' statistics of carbon footprint to date, relies on inadequate measures for sustaining the globe. On the other hand, energy consumption culture has been poorly understood

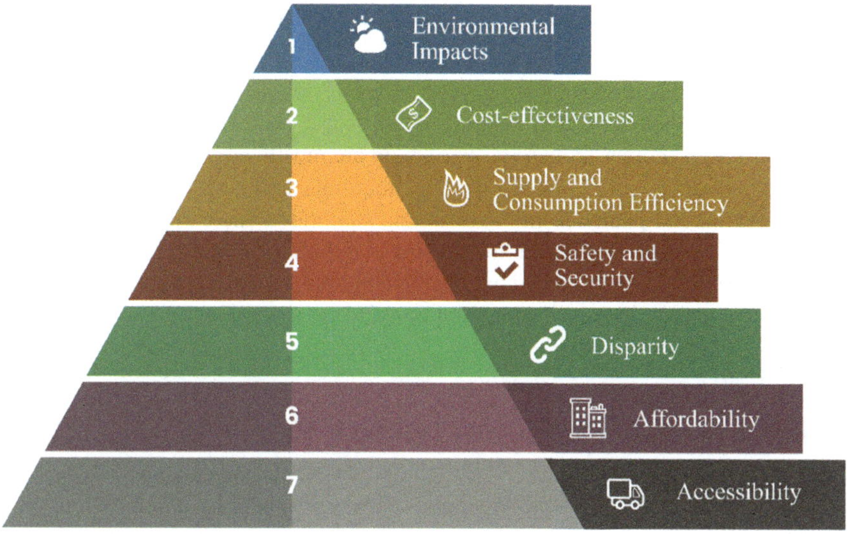

Fig. 1 Economic and technical requirement of environmental-friendly energy provision [16]

in terms of environmental, social, cultural, economic, and psychological variables [13].

However, the recent climate action, the Paris Agreement on Climate Change (2016), as well as the United Nation's Sustainable Development Goals (SDGs) 2030 agenda (2015), are hopefully to be put in action. Regretfully, there have never been full global consensus on realization of such efforts. For instance, withdrawal of the United States of America (USA) from the Paris Agreement [14, 15].

Since energy is an indispensable need of today's life and remains the main element of nations development, provision of environmentally-friendly energy discussed in this study. In addition to the sustainability criteria, some technical and economic requirements are essential for an environmentally-friendly energy provision (Fig. 1).

Reports indicate that still millions of populations in the world do not have access to routine energy services. The world 13% (940 million) populations do not have access to electricity, and 40% (3 billion) populations do not have access to clean fuels for cooking [17]. While imaging a daily life without energy and technology that will lead to go back to past times using primary energy resources with high indoor air pollutions accumulated with a high risk of respiratory diseases [18]. When we come to the context of the least developing countries, access to energy and energy security scenarios become more serious. For instance, the Tokyo blackout in 1987 that affected 2.8 million customers for more than three hours [19].

3 A Brief Overview of Energy Resources and Consumption

Along with the industrial revolution, from about 1760–1840 (or the late eighteenth century), energy has been used as a motive for socio-economic development [20]. Among different sources of energy, six of them are mostly introduced by the human being such as sun (thermal and electric), geothermal (thermal and electric), gravitational potential and planetary motions among sun, moon and earth, chemical reactions, natural resources (thermal and electric), and nuclear energy. The concept of energy consumption and demand are evaluated based on different decision-making models such as switching, selecting, and transferring scenarios.

4 Energy Metrics and Global Interactions

Broadly an energy sector socio-economic analysis complexity is connected to its ethical and behavior demonstration in view of various dimensions with many interactions (local, national, regional, and global. However, since the 1980s combating climate change begun, and in recent years concern of global warming raised. Somehow, it denotes a deficiency in preventive measures in the context of global policies. At the age when the world becomes a small village and human beings have been able to conquer beyond the earth, how it is possible to ignore climate change and global warming. Among many other reasons, the liberalization of energy markets and industrialization can be counted as the main negative contributor to this trend. Globally, greenhouse gas emission has been increased by more than 70% since the 1970s [21]. However, a significant part of this crisis relates to industrial countries, and developing nations owns share in these challenges by having low rate of access to modern energy and the usage of primary source of energy with high indoor and outdoor pollutants.

Due to the importance of energy for nations' development, energy is dealt as a commodity. This establishes nations' economic power and prepares ways for international interactions. Therefore, energy projects at the global scale attract investments. Usually, investments rely on a country tradeability capacity in terms of economic stability and output, balance of budget and payment situations, inflation, interest rate, type of investment, macroeconomic and microeconomic models, and investment decision scenarios. These efforts are based on demand, supply, investment, market, and technical, technological, and institutional opportunities [22].

Apart from pressing need for energy, its ability of storage, transformation and utilization into different forms give it the privilege of international markets [23] that demonstrate a global commodity. Energy resources differences in price (natural gas is more expensive than oil and coal is generally cheap) depends on recovery, storage, and transportation [24].

Gas is a dependable heat source for many consumers worldwide. It is one of the cheap and widely available options for power generation. There are three main types of gas pipelines: gathering lines, which help ferry gas from production wells to pre-processing plants; transmission, which transfers gas from pre-processing plants to distribution systems; and distribution lines. Both the design and operation of gas delivery systems are crucial for high-quality performance. Natural gas transmission systems deliver gas over long distances from producers to consumers. According to Guelpa et al. [25] study, compression is disseminated over the whole natural gas network; it is primarily used to sustain sufficient pressure inside the pipeline to extract gas from the system, which occurs at the city gates as well as the power generators connected to the structure at the transmission level. The main task requires an online declaration, which provides a warranty for economically practical and safe operations; compression's optimized preparation is summarized, minimizing consumption under constrained pressure limits. Optimization procedures are proposed for operations of steady-flow gas pipelines to minimize power usage, with the aim of selecting the best system management [25].

The installation of heat pumps in deliberate locations is undertaken to study the effects of network dynamics and possible critical fluctuations; this should include proper sculpting of both the thermal and electricity networks. Various networks present similarities and discrepancies from a modeling point of view. Gas dynamics are distinct since it is a compressible fluid. The central issue to be modeled is production's randomness, which resolves problems related to disparity and losses. The modeling in a multi-energy structure depends on the control volume measured and the issue to be solved [25].

If there are multiple sources, district heating's (DH) action is managed by variations in mass flow and water temperature produced in the various types of plants. Network models are necessary to optimally incorporate various heat sources and find the most favorable network layout. District heating network modeling is used to analyze the possible modification of the system's performance, profit, and emission; layout changes are utilized, amending operations and unexpected events (e.g., pump failure). Physical models or data interruption can perform modeling [25].

Data interruption or interpolation creates a function through a wide range of experimental data, which can be used in forecasting network behavior in conditions not available in the dataset. It establishes new data points within the range of a discreet set of known data points. A problem that may arise with a data interpolation model is reduced accuracy. When a physical model is used, pressure and temperature distribution along the pipes are permitted. Due to the small density variation of water at temperatures used for heat distribution, the mass flow accumulation can be disregarded. Two scenarios are possible depending upon the requested quantities of pressure, temperature, and network layout (i.e., tree or looped) [25].

The first scenario only requires thermal analysis. If the network does not present loops, only mass and energy conservation equations must be answered; this leads to an easily solved linear problem. The second scenario requires pressure standards. If the network is looped, there is mass flow distribution within the pipelines; deciphering both fluid-dynamic and thermal problems is a priority. In such a situation, the whole set of equations, including the impetus equation, must be answered. This is computationally concentrated, especially in large networks. Various approaches have been proposed to work out the complete set of equations since the DH networks are usually looped in order to better manage possible malfunctions or leakages. Two major aggregation methods have been proposed to reduce the computational costs, which are especially high in the case of large networks. This method's main limitation is associated with the use of transient problems in a looped set of connections. Computational costs can be significantly reduced by solving the various problems at different levels, making the approach suitable for large networks.

First and foremost, distinguish among indicators, indices, and metrics are exigency. Afterward, finding appropriate tools and techniques that can jointly be applied. Finally, used indicators, indices, and metrics categorization, process identification, application, tangible and non-tangible impacts evaluation, and an evolutionary confirmation are essential [26]. According to Waas et al. [6], defining sustainability indicators necessitate some criteria as below:

- Representativeness and interpretation capability
- Simple and easy to interpret
- Scientifically valid
- Able to show trends over time
- Give an early warning and influence about irreversible trends where possible
- Sensitive to change in the environment, society or economy it is meant to indicate
- Based on readily available and adequately documented data
- Capable of being updated at regular intervals
- Have a target level or guideline against which to compare.

As mentioned, access to clean and viable energy requires some criteria for being fit for demands. In the big picture, planning environmentally-friendly energy systems aligned with sustainability and efficiency criteria needs operation reliability, stability, resiliency, efficiency, exergy, and at last overall sustainability within the reduced cost, compressed schedule, optimized resources, mitigated risks, and maintained quality from different perspectives [1].

Eco-exergy is used to estimate the health or sustainability of an ecological system. It is an indicator for evaluating agricultural production from a sustainable point of view, as it is the work energy of the information. Eco-exergy facilitates the environment's policies or strategies. It is used to emphasize the importance of environmental impacts, economic yield in agriculture, natural resources, and economic investment [12]. It is an ideal indicator of ecosystem's development, demonstrating its resilience as it moves toward a higher level of exergy. Resilience to sustainability refers to viable decision-making to offer accessibility, affordability, disparity, safety, use efficiency, supply and production efficiency, cost-effectiveness, and environmental impacts in the best possible way [27]. At all events, using some indicators (composed from different metrics) enable us to measure energy accessibility that generally reflects a comparative and tentative estimation. These indicators can help policymakers and energy sector investors to recognize opportunities, access affectability, evaluate impactability, and propose a solution strategy. Gross Domestic Production (DGP) used as a nation's development index in terms of a function of its total energy consumption as well as carbon monoxide (CO) emissions [28].

Human Development Index (HDI) deals with literacy rate, education level, life expectancy (health care and etc.), and per capita GDP based purchasing power parity of a nation as shown in Eq. (1) [29]:

$$HDI = \frac{EL_I + LE_I + IL_I}{3} \tag{1}$$

- Education level index (ELI)
- Life expectancy index (LE)
- Per capita gross domestic product index based on purchasing power parity, income level index (ILI).

5 Transition Towards Sustainability, Hiring Effective Policies

Environmental sustainability knotted with the main two distinguished topics of renewable energy deployment and energy efficiency. From economic sustainability perspective, both of these options come with a high initial investment with a steadily returning on investment [30]. Understanding of efficiencies (economic efficiency, technical efficiency, technologically efficient, demand efficiency, operation efficiency supply efficiency, decision-making efficiency, etc.) and their alignment with sustainability requirements contribute to socio-economic development and mitigate greenhouse gas emissions [27]. In general, the main challenges towards energy efficiency and energy conservation are [31, 32]:

- Produce, transmit and utilize energy in an environmentally friendly manner.
- Reduce overall costs by enhancement of generation and operation efficiencies.
- Diminish overall costs by the smooth implementation of management and business practices.

Authors in [33] proposed an emerging multi-disciplinaries roadmap for sustainable energy and energy efficiency exploitation in Fig. 2.

Therefore, private and public investment mechanisms are gaining importance in deployment of sustainable energy concepts in terms of renewable resources and technologies with optimum technical and technological efficiencies [34]. By transition from conventional sources of energy (individually use of natural resources in local scale) to renewable technologies (managed-combined environmental-friendly and cost-effectively provision on a national scale) can improve socio-economic and mitigate environmental impacts [35]. A feed-in tariff (FIT) policy is a viable option to put this idea into practice that accelerates investment and attracts investment in the deployment of renewable energy and improvement of energy efficiency. Feed-in-Tariff enables investors to sum up for a long-term purchase agreement for renewable energy production that requires stable access to the grid, stable and long-term energy purchase agreement, and finally prices based on the unit cost of generation [36].

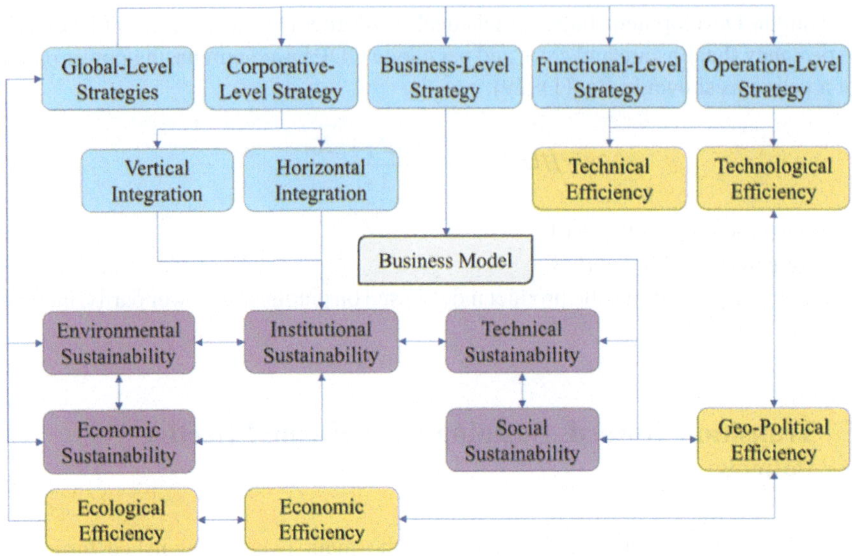

Fig. 2 A multi-disciplinaries roadmap for sustainable energy and energy efficiency deployment [33]

Also, feed-in tariff policies followed by the concept of renewable portfolio standard policy. The renewable portfolio standard policy is quantity-based mechanics standing against the feed-in tariff, which is based on the fixed-price and premium-price [37]. Tax incentive policies is other option that applied in different countries. These tax exemption policies offer fiscal incentive measures to encourage renewable energy deployment that the tax credits could be applied for the investment, production, or consumption segments of electricity generated by renewable energy sources [38].

Ecosystem services further promote environmental sustainability, reinforcing the value of non-monetary ecological qualities and functions. To maintain and advance human well-being, these goals must be met. This can be accomplished by considering ecosystem services as a fundamental component of human well-being [11]. People receive goods and services from the environment and its many ecosystems, such as foods, transportation vehicles, market goods, and other materials or services that humans buy and use. The environment is an asset, generating vital services for all species on earth, both human and nonhuman. Therefore, the sustainability concept is used as a standard for assessing decisions that have environmental consequences.

Industrial ecology is also used to minimize environmental effects by increasing the utilization of energy, capital, and resources; both the production and consumption of goods and services are integrated through design, production, use, and termination [12].

6 Conclusion

Energy remains the primary need and socio-economic motive for nations for decades. That congenially brings welfare and developments to human kinds. At the meantime, global negative competitions and industrialization misdirected this benefit with negative impacts on the climate and environment. Global socio-economic development considering interrelated standardized roadmaps of energy, seems a complicated process. This study addressed sustainability and its pillars in terms of clean and environmentally-friendly solutions. This overview of the subject covers the basics of energy concept to complicated policies for well-management of energy services at different levels. As mentioned, this study offers a concise overview of the main factors influencing energy and environmental sustainability and efficiencies in a systematic manner. Therefore, it can be counted as a compendious reference for students, researchers, scholars, and practitioners related to energy, environment, and sustainable development.

References

1. Danish, M.S.S., Senjyu, T., Funabashia, T., Ahmadi, M., Ibrahimi, A.M., Ohta, R., Rashid Howlader, H.O., Zaheb, H., Sabory, N.R., Sediqi, M.M.: A sustainable microgrid: a sustainability and management-oriented approach. Energy Procedia **159**, 160–167 (2019). https://doi.org/10.1016/j.egypro.2018.12.045
2. Danish, M.S.S., Senjyu, T., Ibrahimi, A.M., Ahmadi, M., Howlader, A.M.: A managed framework for energy-efficient building. J. Build. Eng. **21**, 120–128 (2019). https://doi.org/10.1016/j.jobe.2018.10.013
3. Bhattacharyya, S.C.: Part 1: energy demand analysis and forecasting. In: Energy Economics: Concepts, Issues, Markets and Governance, pp. 9–40. Springer, Berlin (2019)
4. Danish, M.S.S., Senjyu, T.S.: Green building efficiency and sustainability indicators. In: Green Building Management and Smart Automation, pp. 128–145 (2020)
5. Danish, M.S.S., Sabory, N.R., Ershad, A.M., Danish, S.M.S., Ohta, R., Sediqi, M., Ahmadi, M., Senjyu, T.: The least developed countries needs for changing the passive trend of renewable energy exploitation to a proactive trend. Int. J. Energy Power Eng **5**, 215–221 (2016). https://doi.org/10.11648/j.ijepe.20160506.17
6. Waas, T., Hugé, J., Block, T., Wright, T., Benitez-Capistros, F., Verbruggen, A.: Sustainability assessment and indicators: tools in a decision-making strategy for sustainable development. Sustainability **6**, 5512–5534 (2014). https://doi.org/10.3390/su6095512
7. Tiwari, G.N., Mishra, R.K.: Advanced renewable energy sources. Roy. Soc. Chem. (2012)
8. Hill, D.W.: Physics Applied to Anaesthesia. Elsevier (1972)
9. Matin Ibrahimi, A., Masih Sediqi, M., Or Rashid Howlader, H., Sayed Shah Danish, M., Chakraborty, S., Senjyu, T.: Generation expansion planning considering renewable energy integration and optimal unit commitment: a case study of Afghanistan. AIMS Energy 7(4), 441–464 (2019). https://doi.org/10.3934/energy.2019.4.441
10. Danish, M.S.S., Sabory, N.R., Ershad, A.M., Danish, S.M.S., Yona, A., Senjyu, T.: Sustainable architecture and urban planning through exploitation of renewable energy. Int. J. Sustain. Green Energy 6:1–7 (2017). https://doi.org/10.11648/j.ijrse.s.2017060301.11
11. Urbaniec, K., Mikulčić, H., Rosen, M.A., Duić, N.: A holistic approach to sustainable development of energy, water and environment systems. J. Clean. Prod. **155**, 1–11 (2017). https://doi.org/10.1016/j.jclepro.2017.01.119

12. Bilgen, S., Sarıkaya, İ.: Exergy for environment, ecology and sustainable development. Renew. Sustain. Energy Rev. **51**, 1115–1131 (2015). https://doi.org/10.1016/j.rser.2015.07.015
13. Wilk, R.: Consumption, human needs, and global environmental change. Glob. Environ. Change **12**, 5–13 (2002). https://doi.org/10.1016/S0959-3780(01)00028-0
14. Zhang, Y.-X., Chao, Q.-C., Zheng, Q.-H., Huang, L.: The withdrawal of the U.S. from the Paris agreement and its impact on global climate change governance. Adv. Clim. Change Res. **8**, 213–219 (2017). https://doi.org/10.1016/j.accre.2017.08.005
15. Zhang, H.-B., Dai, H.-C., Lai, H.-X., Wang, W.-T.: U.S. withdrawal from the Paris agreement: reasons, impacts, and China's response. Adv. Clim. Change Res. **8**, 220–225 (2017). https://doi.org/10.1016/j.accre.2017.09.002
16. Danish, M.S.S., Senjyu, T., Sabory, N.R., Danish, S.M.S., Ludin, G.A., Noorzad, A.S., Yona, A.: Afghanistan's aspirations for energy independence: water resources and hydropower energy. Renew. Energy **113**, 1276–1287 (2017). https://doi.org/10.1016/j.renene.2017.06.090
17. Ritchie, H., Roser, M.: Access to Energy. Our World in Data. (2019)
18. Sovacool, B.K.: A qualitative factor analysis of renewable energy and sustainable energy for All (SE4ALL) in the Asia-Pacific. Energy Policy **59**, 393–403 (2013). https://doi.org/10.1016/j.enpol.2013.03.051
19. Danish, M.S.S., Senjyu, T., Danish, S.M.S., Sabory, N.R., K, N., Mandal, P.: A recap of voltage stability indices in the past three decades. Energies 12, 1544 (2019). https://doi.org/10.3390/en12081544
20. Beltran, A.: Introduction: energy in history, the history of energy. J. Energy Hist. Revue d'Histoire de l'Énergie 1 (2018)
21. Danish, M.S.S., Sabory, N.R., Danish, S.M.S., Senjyu, T., Ludin, G.A., Noorzad, A.S., Yona, A.: Electricity sector transitions in an after war country: a review of Afghanistan's Electricity. J. Energy Power Eng. **11**, 491–496 (2017). https://doi.org/10.17265/1934-8975/2017.07.008
22. Bhattacharyya, S.C.: Energy economics: concepts, issues, markets and governance. Springer, London (2011). https://doi.org/10.1007/978-0-85729-268-1
23. Thumann, A., Mehta, D.P.: Handbook of Energy Engineering. The Fairmont Press Inc., United States (2008)
24. Fay, J.A., Golomb, D.S.: Energy and the environment: scientific and technological principles. In: Environmental Progress & Sustainable Energy, , 2nd edn, vol. 31, pp. 9–9 (2012). https://doi.org/10.1002/ep.11611
25. Guelpa, E., Bischi, A., Verda, V., Chertkov, M., Lund, H.: Towards future infrastructures for sustainable multi-energy systems: a review. Energy **184**, 2–21 (2019). https://doi.org/10.1016/j.energy.2019.05.057
26. Danish, M.S.S., Senjyu, T., Zaheb, H., Sabory, N.R., Ibrahimi, A.M., Matayoshi, H.: A novel transdisciplinary paradigm for municipal solid waste to energy. J. Clean. Prod. **233**, 880–892 (2019)
27. Danish, M.S.S., Matayoshi, H., Howlader, H.O.R., Chakraborty, S., Mandal, P., Senjyu, T.: Microgrid planning and design: resilience to sustainability. In: 2019 IEEE PES GTD Grand International Conference and Exposition Asia (GTD Asia), pp. 253–258. IEEE, Bangkok, Thailand (2019)
28. Arto, I., Capellán-Pérez, I., Lago, R., Bueno, G., Bermejo, R.: The energy requirements of a developed world. Energy Sustain. Dev. **33**, 1–13 (2016). https://doi.org/10.1016/j.esd.2016.04.001
29. Training Material for Producing National Human Development Reports: United Nations Development Programme. UNDP), New York, USA (2015)
30. Danish, M.S.S., Yona, A., Senjyu, T.: Pre-design and life cycle cost analysis of a hybrid power system for rural and remote communities in Afghanistan. J. Eng. IET **2014**, 438–444 (2014). https://doi.org/10.1049/joe.2014.0172
31. Akadiri, P.O., Chinyio, E.A., Olomolaiye, P.O.: Design of a sustainable building: a conceptual framework for implementing sustainability in the building sector. Buildings **2**, 126–152 (2012). https://doi.org/10.3390/buildings2020126

32. Dahunsi, F.M.: Conceptual framework for sustainable energy development in Africa. In: 2013 IEEE International Conference on Emerging & Sustainable Technologies for Power & ICT in a Developing Society (NIGERCON), pp. 238–241. IEEE, Owerri, Nigeria (2013). https://doi.org/10.1109/NIGERCON.2013.6715661
33. Danish, M.S.S., Elsayed, M.E.L., Ahmadi, M., Senjyu, T., Karimy, H., Zaheb, H.: A strategic-integrated approach for sustainable energy deployment. Energy Rep. **6**, 40–44 (2020). https://doi.org/10.1016/j.egyr.2019.11.039
34. Ye, L.C., Rodrigues, J.F.D., Lin, H.X.: Analysis of feed-in tariff policies for solar photovoltaic in China 2011–2016. Appl. Energy **203**, 496–505 (2017). https://doi.org/10.1016/j.apenergy.2017.06.037
35. Heshmati, A., Abolhosseini, S., Altmann, J.: The energy and environment relationship. In: Heshmati, A., Abolhosseini, S., and Altmann, J. (eds.) The Development of Renewable Energy Sources and its Significance for the Environment, pp. 7–29. Springer, Singapore (2015). https://doi.org/10.1007/978-981-287-462-7_2
36. Couture, T., Gagnon, Y.: An analysis of feed-in tariff remuneration models: implications for renewable energy investment. Energy Policy **38**, 955–965 (2010). https://doi.org/10.1016/j.enpol.2009.10.047
37. Cory, K., Couture, T., Kreycik, C.: Feed-in tariff policy: design, implementation, and RPS policy interactions. National Renewable Energy Laboratory Technical Report NREL. TP-6A2–45549 (2009).
38. Heshmati, A., Abolhosseini, S., Altmann, J.: The Development of Renewable Energy Sources and its Significance for the Environment. Springer, Berlin, Germany (2015)

Aligning Smart City Indicators for Sustainability Outreach: A Case Study

Abdolhamid Ebrahimi, Mujtaba Alemi, Mohammad Qasem Azad, Sayed Hujjatullah Ahmadi, Najib Rahman Sabory, and Mir Sayed Shah Danish

1 Introduction

Daily overcrowding challenges in city management is a big intricate which traditional urban management and development approaches are not sufficient. Due to rapid urbanization, cities lack the mobility, efficiency and capability needed to meet the needs and requirements of citizens for a comfort and secure life. Problems that arise due to urbanization remain unsolved despite enormous scientific advances and achievements. Cities consume 75% of the world's energy and generate a large amount of waste and 70% of greenhouse gas emissions, which contribute to climate change and environmental pollution [1]. On the other hand, this rapid growth of cities stresses urban infrastructure beyond its capacity and capabilities. Therefore, residents and officials suffer from undesirable consequences. Developing countries, such as Afghanistan, are especially under increasing pressure better to provide basic services to a growing urban population. As a result, urban planners around the world seek to integrate all aspects of urbanization by designing models for twenty-first century cities to meet today's new demands and tomorrow challenges. One of the attractive new concepts is the development of smart city, which relies on real-time information on citizens' actions and choices to identify and identify normative and behavioral patterns (both at the city level and the individual level). These data allow officials to understand the behavior of a city over different periods and circumstances and the possibility of influencing them through modeling. Information technology

A. Ebrahimi (✉) · M. Alemi · M. Q. Azad · S. H. Ahmadi · N. R. Sabory · M. S. S. Danish
Kabul University, Kabul 1006, Afghanistan
e-mail: abdolhamid1994ebrahimi@gmail.com

M. Alemi
e-mail: muj.ale.313@gmail.com

M. S. S. Danish
University of the Ryukyus, Okinawa 9030213, Japan

M. S. S. Danish et al. (eds.), *Sustainability Outreach in Developing Countries*,
https://doi.org/10.1007/978-981-15-7179-4_3

can be used as an accelerating factor in improving the quality of life and in achieving the goals of the smart city. Each city in the world has adopted approaches specific to its technological, geographical, political, and social context.

- What are the requirements for Kabul city to be a smart city, and why?
- What are the Smart city indicators for Kabul?

As an emerging term during the 2000s, "smart" has different contexts. In urban policy, "smart" refers to the smart use of information technology to improve the efficiency and effectiveness of urban services and infrastructure [2]. In social science research, it refers to the science of technology in affecting the complex nature of social systems. In engineering, "smart" technology relies on sensor integration for knowledge sharing platforms [3], cloud computing [4], and grid technology for energy management [5].

Brisbane et al. [6]. were considered the first Smart Cities where information and communication technology (ICT) supported social participation, reduced the digital divide, and increased access to services and information. Integration of information systems, services, and urban infrastructure was developed further by technology companies such as Siemens (2004), Cisco (2005), and IBM (2009) to include buildings, transportation, electricity, water and sanitation, security, and health. Currently, many major cities around the world, such as Seoul, New York, Tokyo, Chicago, Amsterdam, Cairo, Dubai, Kochi, Singapore, Trikala (Greece) have launched the smart city Project.

2 Study Objective

Many industrial and international organizations have developed the smart city Agenda, but a review of existing literature shows that there are few domestic studies on the smart city in Afghanistan. Kabul, Afghanistan's capital and largest city, had 5 million people in 2008 and is projected to reach 8 million by 2023 (Central Statistical Office). The high population concentration has led to many complex issues and problems, including.

- Pollution
- Heavy traffic
- Poor urban services
- Urban sprawl.

The centralized, authoritarian planning process and subjective management approach of city administration have contributed to a lack of attention to the city's real needs, and a lack of accountability to citizens has further caused instability. Smart city theory and cyberspace offer real-world solutions to alleviate city problems to enhance the quality of life of Kabul's citizens and the city's international reputation. The purpose of this article is to investigate the indicators and the requirements of

Kabul city towards smart city standards. A clear understanding of the actual characteristics and requirements of each city (reflecting its conditions and characteristics) prevents ineffective use of limited resources and promotes adopting appropriate strategies and actions.

This descriptive-analytical study relied on document, library, and field information obtained from two closed-ended questionnaires (Likert's five-choice range). The first questionnaire had 30 items (each with 23 questions), which focused on requirements for smart city and was purposefully sampled by people with sufficient knowledge of urban issues. The second questionnaire surveyed smart city indicators with 100 items (each with 36 questions) and was administered through random sampling. Data from the questionnaire were analyzed with Microsoft Excel.

- Expressing the requirements for moving toward smart city
- Evaluating smart city Indicators in Kabul.

3 Smart City Requirements

A concept that has received a lot of attention in recent years in urban planning, smart city focuses on three main areas: academic, industrial, and governance [7]. First, the academic literature suggests improving three areas of governance, social development, and the environment. From the industrial perspective, smart cities acknowledge the interaction between competition and sustainable development by identifying sustainable productivity, environment, and social development as goals. Finally, the government perspective has focused more on international challenges, including quality of life, economic growth, environment, energy, sustainability, security, health, and mobility.

An in-depth analysis of literature shows a common emphasis on smart city concepts. Specifically utilizing information and communication technology in urban services and infrastructure, integrating different systems in planning and implementation, and building cooperation among stakeholders in all stages of urban development, learning, and management of local resources. Here are some definitions:

- A smart city combines physical, information technology, social, and business infrastructures to enhance the community [8].
- Smart cities are high-capacity learning and innovation territories based on the creativity of citizens, institutions, knowledge-based organizations, and their digital infrastructures for communication and knowledge management [9].
- Smart cities are the result of creative and knowledge-based strategies that enhance the competitive, supportive, ecological, socio-economic performance of cities. Such smart cities are based on a combination of promising human capital (skilled labor, infrastructure capital), high-tech communication facilities (social capital), open and intense network communications (entrepreneurial capital), and creative/risky business activities [10].

- Smart cities are highly productive as they have a high proportion of people with higher education, knowledge-based jobs, output-oriented planning systems, creative activities, and sustainable orientation initiatives [11].
- A smart city refers to a local entity, department, city, region or small town that adopts a holistic approach to real-time information technology analysis and encourages sustainable economic development.
- Smart city is a society with a medium level of technology, interconnectedness, sustainability, comfort, attractiveness, and safety [12].
- The use of information and communication technology (ICT) with its effects on human capital, communication capital, and social and environmental issues, often illustrated by the concept of smart cities [13].
- A smart city informs its physical infrastructure to improve comfort, facilitate movement, increase efficiency, conserve energy, enhance the quality of climate, identify and resolve problems, speedily recover after an emergency, collect data for better decision making, deploy resources efficiently, and enable data-sharing and collaboration between entities and sectors [14].
- Smart or creative city experience fosters an innovative economy by investing in a quality of life that attracts highly educated citizens to live and work in the smart city [15].

3.1 Case Study

The capital and largest city of Afghanistan (Fig. 1), Kabul has seen rapid population growth and urbanization over time, with the population of Kabul in 1925 at over 90,000. The development of roads and relief areas in the 1940s and 1950s resulted in a population increase to 380,000 in 1962. The first Kabul Master Plan prepared by the Afghan and Soviet missions between 1962 and 1964 included 800,000 people. At the beginning of the invasion of the former Soviet Union, the population of Kabul in 1979 was over 1.5 million. After the fall of the Taliban in 1999, many refugees returned home, bringing Kabul's population to 1.78 million. The Central Statistics Office (CSO) [16] announced the 2005 population of Kabul at 2.26 million within the 14 central districts and 2.72 million taking all 22 districts into account. By 2008, the population of Kabul was 4.2 million, and the population is estimated to reach 8 million by 2023.

3.2 Investigation of Smart City Strategies in Kabul City Master Plans

Two master plans were prepared for Kabul city after the civil wars: JICA in 2011 and Sasaki in 2017. We decided to investigate how these master plans include strategies that help Kabul progress to Smart city standards.

Fig. 1 Map of Kabul city [17]

3.3 *Sanitation Infrastructure for Removing Sludge and Wastewater Treatment*

Sanitation infrastructure includes solid waste and water management facilities. Wastewater is treated by a small decentralized wastewater treatment systems plant and released into a constructed wetland, helping to restore the land and providing a landscape amenity (Table 1).

Table 1 Prerequisites and public benefits of sanitation infrastructure [18]

Prerequisites	Public benefits
Decentralized wastewater treatment system and constructed wetlands	Access for Kabul citizens to sanitation facilities and improvement of their local environments
Material recovery facility	Restoration of lands
Sanitary sewers to carry sludge to decentralized wastewater treatment systems	Jobs and income generation
	A model for community facilities that can be replicated across Kabul

3.4 Solid Waste Infrastructure

While the JICA plan proposed 6 locations for landfills, only three were practical: Pulicharki (east), Childokhataran (south), and Morghgiran (west). Four sites for transfer stations were also proposed. JICA proposed more containers (380 m^3 capacity) and collection vehicles (180 m^3 capacity) for solid waste collection and other facilities and urged special treatment for solid medical waste.

3.5 Pollution Control

The Sasaki master plan has proposed three approaches for waste collection [19]:

- Residential low rise and organic neighborhoods: Residents sort their waste at home and transport it to their nearest community collection points.
- High-rise residential and mix use: Commercial and high-rise residential buildings have loading docks for solid waste collection within the building.
- Industrial and institutional parcels: Industrial areas and large parcels have their own solid waste collection management loading zones.

3.6 Wi-Fi in Major Public Spaces

Partners with local internet offer free public Wi-Fi in major public spaces, including the community library. Free public Wi-Fi provides vital access to information for students, professionals, and everyday citizens, particularly those who cannot afford to access the internet at home. Public Wi-Fi can transform underutilized public spaces into hubs of activity that supports retail, food, and other business opportunities.

3.7 Economy

Special Economic Zones: Areas where trade and business laws differ from the rest of the country to attract international investment, creating jobs, and better integrating with the global economy. Create vocational and technical institutes and schools: These schools provide an alternative to traditional universities for students with limited time and resources and can help grow targeted sectors of a national or regional economy.

3.8 Mobility

JICA has proposed BRT and LRT networks to connect and integrate urbanized areas of Kabul highest demands, both for congestion and efficiency. Auxiliary roads connect the main arterial and ring roads throughout Kabul.

3.9 Water Supply

According to MUDH [20], by 2025, the population of the entire Afghanistan Urban Water Supply and Sewerage Corporation (AUWSSC) service area is projected to be 6.2 million. A critical component in the drinking water supply plan is the immediate construction of the Shatoot Dam, which will provide 87.2 Mm^3/year of domestic water supply. Local groundwater resources would provide 33.2 Mm^3/year, enough to meet remaining demand. Industry needs (9.1 Mm^3/year) would be drawn from the Kabul River at a point east of central Kabul and treated at a facility. The planned drinking water distribution network consisted of both house connections and public taps.

3.10 Renewable Energy Projects in Kabul

The JICA Kabul Master Plan has listed these renewable projects with their capacity and costs for construction (Table 2):

4 SDG's Requirements for Safe, Sustainable and Inclusive City

By adopting the smart city model, SDG's goals for a sustainable, safe, and a comprehensive city can be attained (Table 3):

5 Result and Discussion

5.1 Evaluation of Smart City Requirements for Kabul

The most common global challenges that motivate cities towards smartness are listed below:

Table 2 Renewable projects projection in Kabul city by 2025 [18]

Project name	Capacity (MW)	Estimated cost (million USD)	Start year	Estimated completion	Remarks
Naghlu solar project	100	180	2017/2018	–	Feasibility of study has been completed
Kabul solar rooftop project	10	20	2017/2018	–	Solar data are available
Sha Wa Aroos hydro project	1.2	48	Started	2018	In progress
Sarubi 2 hydro project	180	700	2021/2022	–	Feasibility of study has been completed (1980)
Chak Wardak	9.3	13.2	Started	2019	Recently contracted
Parwan wind project	10	25	Started	2017	Data have been collected and analyzed
Baghdara Kapisa	240	560	2016/2017	2019/2020	Feasibility of study has been completed

- Rapid urbanization
- The impact of cities on the environment
- Climate change and greenhouse gas emissions
- Economic crises
- Demographic changes
- Other factors such as (technological advances, ICT, brain migration, administrative bureaucracy, transportation problems, water, energy, etc.).

Urban planning experts completed a questionnaire on the sub-criteria for each of the requirements for Kabul city to become smart (Fig. 2) using a Likert five-range scale. A factor's mean value that is larger than 3 and closer to 5 expresses the high importance of that factor; a value smaller than the number 3 and closest to the number 1 indicates its low impact. Of the highest importance was environment factor with an average of 4.33. "Other factors" had the second-highest average (4.17) and included concerns such as bureaucracy, ICT advances, crime prevention, and wasteful use of critical resources. The remaining factors by order of importance were rapid urbanization (4.16), economic factors (4.00), demographic factors (3.91), and education.

Table 3 The SDGs requirements for safe, sustainable, and inclusive cities [21]

General	Infrastructure	Environmental Sustainability	Governance and Legislation
Proportion of population living in households with access to basic services	Traffic facilities	Population exposed to outdoor air pollution	Investment capacity
Sustainable and resilient buildings	Access to improve water	Wastewater treatment	Local expenditure efficiency
Proportion of wastewater safely treated	Access to improved sanitation	Share of renewable energy	Public–private partnership
Accessible and inclusive transport system	Access to electricity	Solid waste recycling share	
Participatory and inclusive urbanization	Mobile network coverage		
Access to safe and inclusive public space	Internet access		
Proportion of total adult population with secure tenure rights to land, with legally recognized documentation, and who perceive their rights to land as secure, by sex and length of tenure			

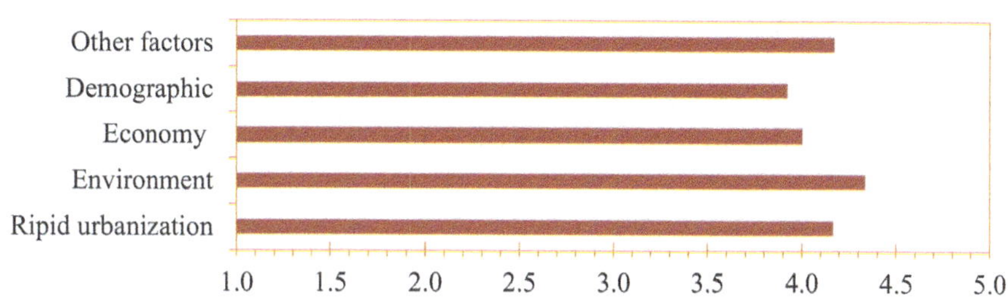

Fig. 2 Importance of smart city requirements in Kabul city

5.2 *Environment*

Among the environment sub-criteria for Kabul as a smart city, sewer and groundwater management was the most significant concern, confirming that the lack of a sewage system remains a high priority. Climate change and increasing greenhouse gases were the lowest priority among these sub-criteria (Fig. 3).

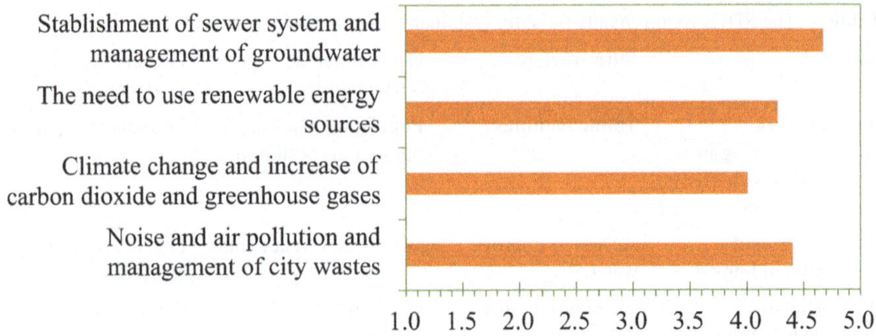

Fig. 3 Importance of environmental sub-criteria

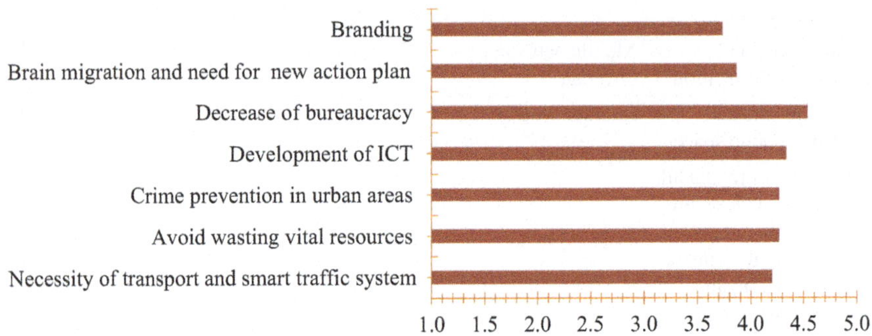

Fig. 4 Importance of "other factors" sub-criteria

5.3 *Bureaucracy*

Among the "other factors" sub-criteria for smart city planning, bureaucracy was the highest concern (4.53), with improvements in ICT (4.33) and urban crime prevention (4.27) following in importance (Fig. 4).

5.4 *Rapid Urbanization*

Among the "urbanization" sub-criteria towards smart city planning, having a progressive local strategy to balance horizontal and vertical city development had the highest score: 4.47 (Fig. 5).

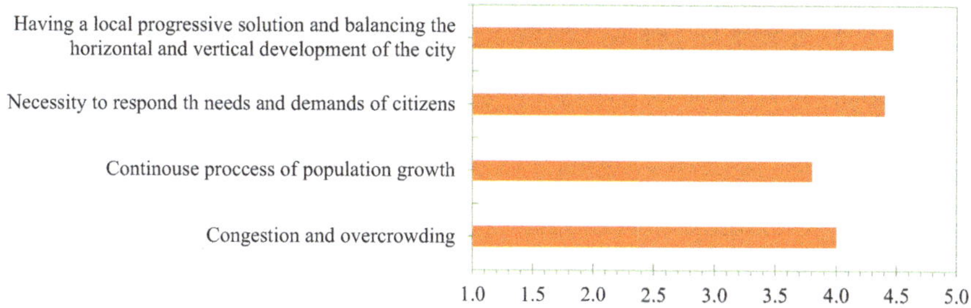

Fig. 5 Importance of rapid urbanization factor sub-criteria

Fig. 6 Importance of economic sub-criteria

5.5 *Economy*

Among the "economic factors" sub-criteria towards smart city planning, attracting a skilled and creative workforce was most important: 4.13 (Fig. 6).

5.6 *Demographic*

Among the "demographic factors" sub-criteria towards smart city planning, supporting public education was the highest priority: 4.40 (Fig. 7).

6 Evaluation of Smart City Indicators in Kabul

From our literature search to identify smart city indicators, six common indicators emerged: smart mobility, smart people, smart living, smart environment, smart governance, and smart economy [22]. A 36-item questionnaire was developed, six per indicator domain, according to Table 4.

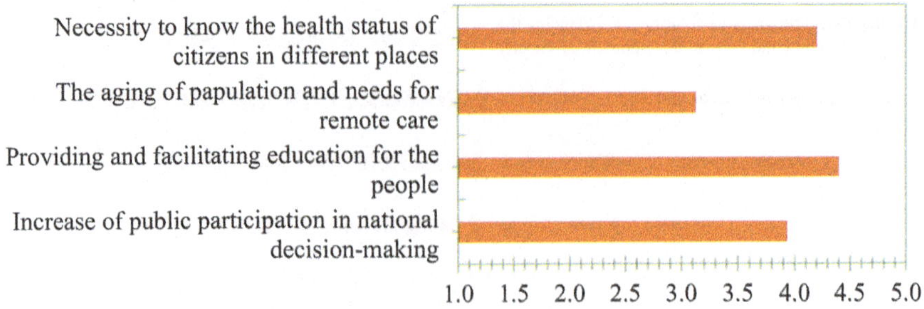

Fig. 7 Importance of demographic sub-criteria

Table 4 Smart city indicators [22]

Smart economy	Smart citizen	Smart governance
– Innovative spirit	– Level of qualification	– Participation in decision-making
– Entrepreneurship	– Affinity to lifelong learning	– Public and social services
– Economic image and trademarks	– Social and Ethnic plurality	– Transparent governance
– Productivity	– Flexibility	– Political strategies and perspectives
– Flexibility of labor market	– Creativity	
– International embeddedness	– Cosmopolitanism	
– Ability to transform	– Participation in public life	
Smart mobility	Smart environment	Smart living
– Local accessibility	– Attractively of natural conditions	– Cultural facilities
– (Inter-)national accessibility	– Pollution	– Health conditions
– Availability of ICT infrastructure	– Environmental protection	– Individual safety
– Sustainable, innovative and safe transport system	– Sustainable resource management	– Housing quality
		– Education facilities
		– Touristic attractively
		– Social cohesion

Based on the questionnaire results for the indicators of smart city in Kabul, the respondents consider Kabul city in "critical condition" in achieving smartness criteria (Fig. 8).

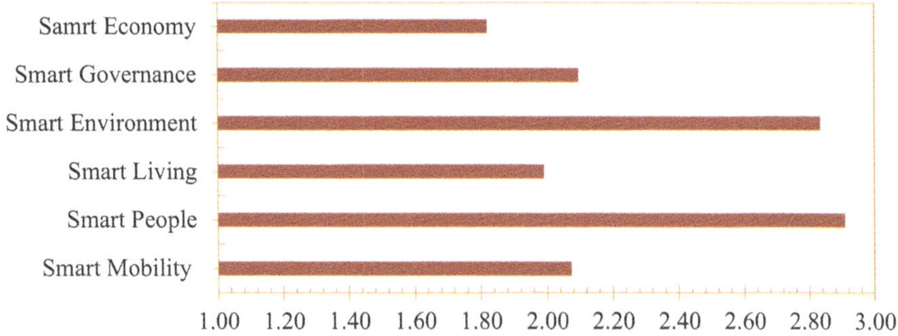

Fig. 8 Status of Smart city indicators in Kabul city

7 Recommendations

Based on these findings, Kabul should make progress towards smart city standards to address its current and future problems. To improve the status and level of smart city indicators in Kabul city, a series of strategies and actions must be taken in parallel by the government, the private sector, and ultimately the people. Our recommendations contains prerequisites and contexts that must be provided before addressing practical solutions.

7.1 Prerequisites and Platforms

- Integrated policies: Kabul's abrupt political changes have created confusing and conflicting policies affecting urban affairs. Reaching the smart city standards requires an evolutionary, continuous process, and government and municipal authorities should adopt coordinated, integrated, and independent from policy changes to prevent further redundancy or stoppage of urban policies in the long run.
- Legislation: Discussion on these issues among government, people, and the private sector requires trust, transparency, and promotion of justice through widespread and equal access to information. Open collaboration fosters a good learning environment and enhances the collective intelligence of cities. Legal and regulatory barriers must be corrected or removed. Confidentiality, privacy, and intellectual property rights should also be protected by governments. Citizenship rights tailored to the needs of the information society should define the boundaries in the private and public spheres.
- Integrated vision: A smart city vision takes advantage of the creativity and synergy among different communities and stakeholders, increasing the capacity for innovation, organizational change, and equitable dissemination of knowledge. For

example, transitional change of government should be achieved according to the expectations of the citizens consistent with this unified vision.
- Security: Investment in security resources will prevent or mitigate societal factors that result in decreased financial investment, "brain drain", infrastructure destruction, economic or trade imbalances, or the inefficient or wasteful use of our talent or natural resources.

7.2 Practical Solutions

Table 5 presents some possible practical solutions that can build on the prerequisites to bring Kabul closer to the ideals of a smart city.

Table 5 Possible actions and strategies to address smart city indicators for Kabul city

Possible actions and strategies	Indicators
Smart citizens	– Increasing Internet access for citizens – Facilitate education for people – Increased convenience for those using ICT tools – Support for educational institutions – A foundation for e-learning – Increasing remote learning
Smart living	– Using ICT in health services – Creating social and personal security – Investing in cultural sectors – Emerging infrastructures for emergency services – Providing social justice – Investing in social health
Smart government	– Use of ICT to reduce bureaucracy – Increasing transparency in governance – Increasing citizens' involvement in decision-making – Create public access to official documents – Elections and election participation – Establishing electronic office systems
Smart environment	– Pollution control – Creating city-wide canalization – Groundwater recharge – Implementation of renewable energy projects – Licensing for green buildings – Create and encourage hiking and bike lanes
Smart economy	– Connection with global markets – Electronic banking – Ease of investment for the private sector – Discovering natural resources – Investment in agriculture and livestock – Attract international investors

(continued)

Table 5 (continued)

Possible actions and strategies	Indicators
Smart mobility	– Increase energy-efficient cars and fuels – Creation of electric transport systems – Increased transportation systems – Increase bike lanes – Creating an intelligent system for traffic control – Improve basic transportation infrastructure

8 Comparison

We can compare the present and proposed/future state of Kabul city in Table 6.

Table 6 Comparison of present and future

Indicators	Future	Present
Smart citizens	– Low level of qualification – No interest in lifelong learning – Lack of flexibility – Weak creativity – No cosmopolitanism – Participation of individuals in public life	– High level of qualification – Great investment in lifelong learning – Flexibility – Creativity – Cosmopolitanism – Participation of individuals in public life
Smart Living	– Lack of cultural facilities – Bad health conditions – Low level of safety – Bad housing quality – Weak education facilities – Bad situation for tourism	– Increase in cultural facilities – Better health conditions – High level of safety – Housing quality – Education facilities – Attraction of tourists
Smart government	– Self-decision-making – Not fair public and social services – Opaque governance – Weak political strategies and perspectives	– Participation in decision-making – Good level of public and social services – Transparent governance – Smart political strategies and perspectives
Smart environment	– Bad condition of natural conditions – Too much pollution – Environmental destruction – Waste of resources	– Attractive natural conditions – Lowest level of pollution – Environmental protection – Better management of Sustainable resources
Smart economy	– Lack of innovative spirit – No entrepreneurship – Low level of productivity – Weak condition of labor market	– Innovative spirit – Entrepreneurship – Productivity – Flexibility of labor market

(continued)

Table 6 (continued)

Indicators	Future	Present
Smart mobility	– Weak condition of local accessibility – Weak (Inter-)national accessibility – Lack of ICT infrastructure – Unfair condition of sustainable, innovative and safe transport system	– Increase of local accessibility – Developed (Inter-) national accessibility – Availability of ICT infrastructure – Sustainable, innovative and safe transport system

9 Conclusion

Rapid expansion of cities in developing countries such as Afghanistan has caused many problems with environmental degradation, poor infrastructure effectiveness, pollution, high population density, decreasing quality and availability of urban services, increased intercity travel time and distance, increased energy consumption, and the loss of human, social, and natural capital. In response, new urban planning models such as "smart city" were proposed. A smart city relies on responsive-smart infrastructure and communication technologies as a critical solution. Literature reviews and theory guided this examination of the requirements and smart city indicators to help Kabul become a smart city. Environment requirements were thought to be of the highest importance, but all smart city indicators are considered to be in critical condition.

References

1. Ferraro, S.: Smart Cities: Analysis of a Strategic Plan (2013). https://amslaurea.unibo.it/5420/1/Ferraro_Saverio_tesi.pdf
2. Karadag, T.: An Evaluation of the Smart City Approach (2013). https://etd.lib.metu.edu.tr/upload/12615687/index.pdf
3. Mancilla-Amaya, L., Sanín, C., Szczerbicki, E.: Smart knowledge-sharing platform for E-decisional community. Cybern. Syst. **41**, 17–30 (2010). https://doi.org/10.1080/01969720903408730
4. Kim, S., Song, S.M., Yoon, Y.I.: Smart learning services based on smart cloud computing. Sensors (Basel) **11**, 7835–7850 (2011). https://doi.org/10.3390/s110807835
5. Arulmurugan, V.S., Vijayan, S.: Quality of experienced based approach for power scheduling in smart grids. Life Sci. J. **10**, 1724–1728 (2013)
6. Alvarez, F., Cleary, F., Daras, P., Domingue, J., Galis, A., Garcia, A., Gavras, A., Karnourskos, S., Krco, S., Li, M.S., Lotz, V., Müller, H., Salvadori, E., Sassen, A.M., Schaffers, H., Stiller, B., Tselentis, G., Turkama, P., Zahariadis, T.: The Future Internet. Springer, Berlin, Germany (2012)
7. Mosannenzadeh, F., Vettorato, D.: Defining smart city. A conceptual framework based on keyword analysis. TeMA J. Land Use Mobil. Environ. 683–694 (2014). https://doi.org/10.6092/1970-9870/2523
8. Harrison, C., Donnelly, I.A.: A theory of smart cities. In: Proceedings of the 55th Annual Meeting of the International Society for the Systems Sciences, vol. 55, pp. 1–15 (2011)

9. Schaffers, H., Komninos, N., Pallot, M., Trousse, B., Nilsson, M., Oliveira, A.: Smart cities and the future internet: towards cooperation frameworks for open innovation. In: Domingue, J., Galis, A., Gavras, A., Zahariadis, T., Lambert, D., Cleary, F., Daras, P., Krco, S., Müller, H., Li, M.-S., Schaffers, H., Lotz, V., Alvarez, F., Stiller, B., Karnouskos, S., Avessta, S., Nilsson, M. (eds.) The Future Internet, pp. 431–446. Springer, Berlin, Germany (2011). https://doi.org/10.1007/978-3-642-20898-0_31
10. Kourtit, K., Nijkamp, P.: Smart cities in the innovation age. Innov. Eur. J. Soc. Sci. Res. **25**, 93–95 (2012). https://doi.org/10.1080/13511610.2012.660331
11. Kourtit, K., Nijkamp, P., Arribas, D.: Smart cities in perspective—a comparative European study by means of self-organizing maps. Innov. Eur. J. Soc. Sci. Res. **25**, 229–246 (2012). https://doi.org/10.1080/13511610.2012.660330
12. Lazaroiu, G.C., Roscia, M.: Definition methodology for the smart cities model. Energy **47**, 326–332 (2012)
13. Lombardi, P., Giordano, S., Farouh, H., Yousef, W.: Modelling the smart city performance. Innov. Eur. J. Soc. Sci. Res. **25**, 137–149 (2012). https://doi.org/10.1080/13511610.2012.660325
14. Nam, T., Pardo, T.A.: Conceptualizing smart city with dimensions of technology, people, and institutions. In: Proceedings of the 12th Annual International Digital Government Research Conference on Digital Government Innovation in Challenging Times—dg.o '11, p. 282. ACM Press, College Park, Maryland (2011). https://doi.org/10.1145/2037556.2037602
15. Thite, M.: Smart cities: implications of urban planning for human resource development. Hum. Resour. Dev. Int. **14**, 623–631 (2011). https://doi.org/10.1080/13678868.2011.618349
16. Central Statistics Organization (CSO)—Afghanistan: Statistical indicators in Kabul province. Central Statistics Organization (CSO)—Afghanistan, Kabul, Afghanistan (2019)
17. Kabul City Map (2019)
18. Draft Kabul City Master Plan: Product of Technical Cooperation Project for Promotion of Kabul Metropolitan Area Development Sub Project for Revise the Kabul City Master Plan (2011). https://openjicareport.jica.go.jp/pdf/12058566_01.pdf
19. Sasaki: Kabul urban design framework. Ministry of Urban Development and Housing, Kabul, Afghanistan (2017)
20. Ministry of Urban Development and Housing (MUDH)—Afghanistan: Shatoot Dam Layout (2019)
21. United Nations (UN): Sustainable Development Goals (SDGs). https://sustainabledevelopment.un.org/sdgs. Last accessed 2019/11/01
22. Correia, L.M., Wünstel, K.: Smart Cities Applications and Requirements (2011)

Optimal Merging of Transportation System Using Renewable Energy-Based Supply for Sustainable Development

Mikaeel Ahmadi, Mir Sayed Shah Danish, Tomonobu Senjyu, Habibullah Fedayee, Najib Rahman Sabory, and Atsushi Yona

1 Background and Introduction

Today's world vision towards energy is favored by many sectors and several perspectives. Energy has been used since human creation and the progressive need to impact and influence human life led to various approaches as well as different resources discovered by humanity. Nowadays, fast development in technologies and discovery creates huge gaps that require a wider area of energy resources deployment. The power system and transportation are the two main focused sectors worldwide. Electrification in transportation or electric mobility, particularly battery-based electric vehicles (EV) as private and public transport, is changing not only the world of the transportation sector but also the impact on power system behavior. The power system and transportation sectors are merging via EVs deployment. EVs and electric buses (EBs) are getting significant attention worldwide due to their performance and fast growth in the transportation sector in many megacities. Zero emission electric vehicles are powered by electricity, which can be produced from clean energy, such as solar, wind, and water energy, to mainly improve environmental friendly and oil shortage issues. These vehicles can also serve as distributed energy storage units, used to balance fluctuations of the power system. There are some major issues in both sectors, as highlighted in Fig. 1, that optimal deployment of recent battery electric mobility can contribute to overcoming the raised issues partially.

At the end of 2017, Shenzhen, a city in southern China with 12.5 million population, was the first megacity in the world realizing total bus electrification. In China, many other cities are following the same strategy towards clean energy-based public

M. Ahmadi (✉) · M. S. S. Danish · T. Senjyu · H. Fedayee · A. Yona
University of the Ryukyus, Okinawa 903-0125, Japan
e-mail: ahmadi.micaeil@gmail.com

N. R. Sabory
Kabul University, Kabul 1006, Afghanistan

M. S. S. Danish et al. (eds.), *Sustainability Outreach in Developing Countries*,
https://doi.org/10.1007/978-981-15-7179-4_4

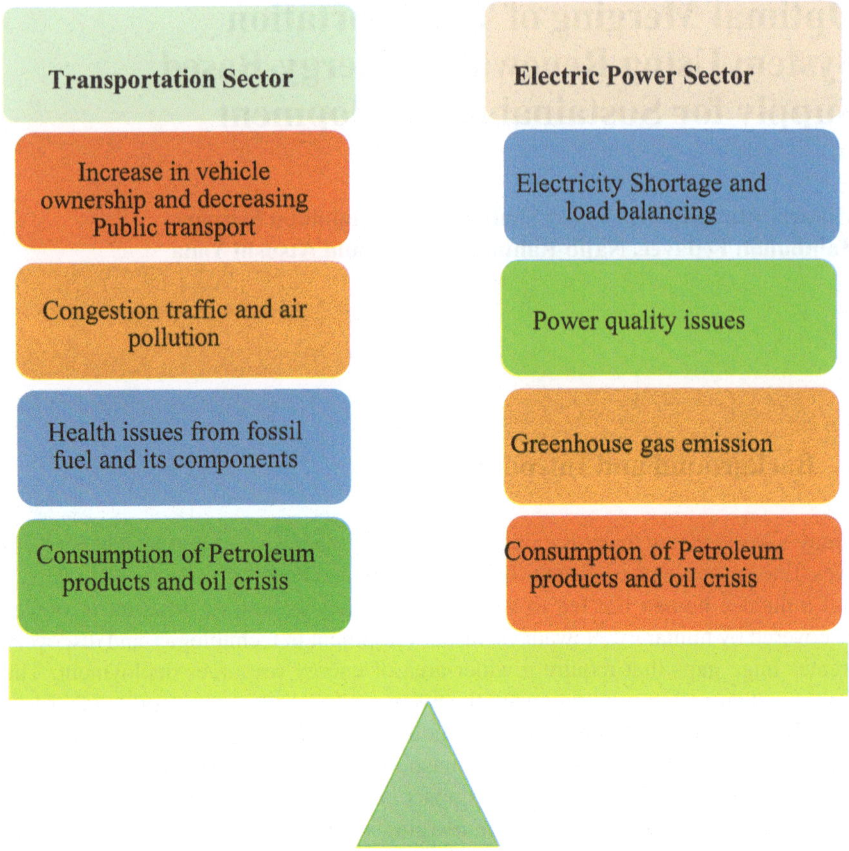

Fig. 1 Major issues in transport and electric power sectors

transportation, and likewise, many countries around the world commit to turning to electric public transportation [1]. The charging infrastructures beside full battery electric bus are very important for welcoming this technology in cities and require a much higher investment than traditional bus fleets. Plug-in charging mode is divided into two types of charging stations: slow charging and fast charging. Nowadays, plug-in charging technology dominates the e-bus market. While swapping stations are said to be challenging due to battery security risks and instability of vehicles chassis by frequent swapping. In Shenzhen all 16,359 e-buses have utilized the plug-in fast charging mode for their power supply. The charging stations can provide night parking, charging maintenance, and logistics services for the e-buses. The first large-scale bus charging station, Yueliangwan station, has 11 floors with the building area of 98,478 m^2 covering 660 e-bus parking spots and 237 chargers installed [2, 3].

To introduce plug-in charging stations for the public governmental/private sector facilitated with EBs to grid (EB2G) incorporating renewable power stations in order to improve the power supply and quality during park state is an attractive research.

This multitarget aspect is very vital for developed and developing countries which creates an interactive platform in the related sectors.

Therefore, in planning stages it can be deeply analyzed to consider the extent of its contribution economically, technically, and environmentally. Since the power and transportation sectors parallelly can get benefit from technical and economic points of view, the investment cost for the implementation of such projects can be divided upon related sectors and thus be affordable. The charging stations can be counted as coupling points which can interact as supply chain by enabling EB2G option. Considering this option, the power sector supplies energy for battery EBs to meet their requirement of traveling distances according to predetermined schedules, and in return, during the parking stage based on EB's state of charge and a smart control mechanism, it can supply the grid. Therefore, it is worthy of mentioning that the appropriate selection of location and sizing of charging stations can significantly impact on reducing the cost related to the power grid.

2 Literature Survey

The integration of electric vehicles into the power system, especially distribution networks, is a hot topic among researchers nowadays, and many research studies are conducted around the world. However, very few articles on the influence and impact of electric mobility in public transportation have been published so far, which indicates inadequate research in this area. For this reason, it can be said that more extensive research is also needed to help advance this technology.

A multiobjective optimization based on single objective GA considering different weight for objective functions is conducted for the purpose of network reconfiguration. EVs with enabled V2G option along with access to renewable energies are introduced to affect the voltage and loss indexes as well as improving the economy and reliability of the proposed test distribution network. However, no optimal sizing and placement is applied nor investigated to fulfill the lack of methodology validation [4].

Reference [5] conducted a research review on the recent three decades of how to mitigate unbalanced scenarios in power systems using network reconfiguration. The uncertainty of EV users, intermittency of renewable energy resources, and frequently changing single-phase loads have been investigated based on several attempts using different optimization tools.

Authors in [6] developed and applied a robust multiobjective optimization scheduling model to solve the proposed problem of large-scale EV penetration under uncertainty. Multiple scenarios are introduced and solved considering the concept of economic analysis. Wind power as an intermittent renewable energy resource is utilized along with adjustment of SOC of EVs to fulfill the fluctuation of the wind power.

Recent and interesting research on optimization of shared autonomous EVs (SAEVs) in a virtual power plant (VPP) is deployed including renewable-based microgrid [7]. All historic data utilized in the simulation are from Japan-related agencies and the application of this technology is promising for the next decade which is showing 20% lower cost compared to the current private EVs utilization.

3 Contribution Summary

All previous studies are conducted to analyze the EVs concern in the power system. However, few studies are deployed for consideration of electric mobility in public transportation. In this study, electric buses are focused on because of their more practicability in implementation and scheduling. In this study, the importance and optimal deployment of battery EBs are investigated to highlight the advantages of this technology in all-inclusive aspects. Moreover, this study attempts to classify the related sustainable development goals (SDGs) and highlight the contribution of the proposed approach to meet related SDGs.

Opportunities and challenges in transportation and power sectors particularly in developing nations such as Afghanistan are further explored in this research to act firstly as worthy awareness to policy makers and related authorities and secondly seeks to highlight the all-inclusive benefit of electric public transportation as somehow opposed to some of the existing thoughts. Furthermore, a model is proposed that optimally incorporates EBs charging stations, rooftop solar PV, and storage system into a test distribution network. A population-based multiobjective optimization algorithm with an inner power flow method is simulated and utilized to conduct the proposed approach in time based and one-hour horizon.

4 Complying SDGs Related to Electricity and Transport Sector

The 2030 Agenda for Sustainable Development, implemented by all United Nations Member States in 2015, provides a shared blueprint for peace and prosperity for people and the planet, currently and into the future. There are 17 SDGs, which are an urgent call for action by both developed and developing nations in a universal corporation [8]. Power and transportation sectors are two important sectors that comply to target goals of the sustainable development goals (SDGs) directly and indirectly. Although each of the SDGs and its related targets can be extensively influenced by every one of these sectors, five SDGs in common, which power system and transportation significantly contribute to their targets, are highlighted in detail (Fig. 2).

- Goal 7. Affordable and clean energy: Power systems and transportation sectors are moving towards the utilization of renewable energy and smart technologies which automatically improves the targets of the seventh SDG to ensure access

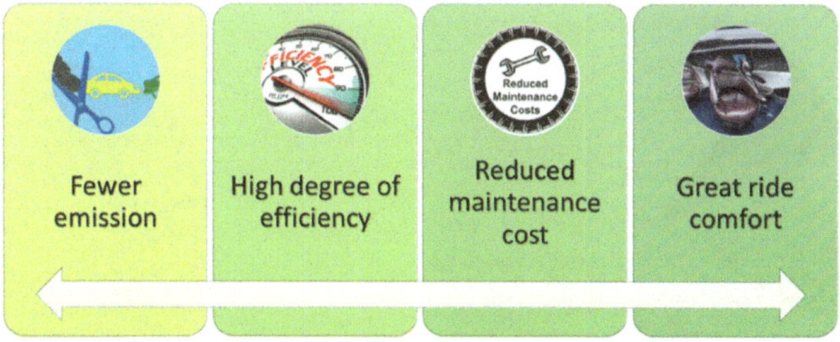

Fig. 2 All-inclusive benefits of electric public transportation

to affordable, reliable, sustainable, and modern energy for all. Since the focus of this paper is in the direction of merging power systems and public transportation in the means of battery electric public mobility, the penetration of clean energy is highly demanded.

- Goal 9. Industry, innovation, and infrastructure: In developing countries such as Afghanistan, power systems and transportation infrastructures are yet unplanned or in planning and building stages. The proposed approach as an optimal merging of the power system and transportation yields to expected targets of this SDG in building resilient infrastructure, promoting inclusive and sustainable industrialization, and fostering innovation.
- Goal 11. Sustainable cities and communities: To support this SDG, developing sustainable public transportation and investing in renewable energy resources to reduce the negative environmental impact by cities are two substantial issues. Furthermore, incorporating relevant technologies such as storage based electric mass transportation and enabling V2G options significantly contribute to attaining the targets of the eleventh SDG as well as acting as a support to the electricity grid from dependency to import power.
- Goal 12. Sustainable consumption and production: Restraining the usage of natural resources, particularly nonrenewable energy resources in a sensible manner enhances sustainability from production to consumption by efficient preservation. Constructive decisions based on policy and strategies can let the mobility as a focus of sustainable consumption into an efficient path of mutual contribution with electricity production. Through this approach, a sustainable pattern of consumption and production is auspicious.
- Goal 13. Climate actions: Climate changes and its adverse effects not only on human beings but every living thing requires urgent actions combating this man-caused disaster. These environmental changes have prompted human beings to take action and not take any measures that are effective in deteriorating the living environment. Transportation and power systems as the highest energy consumer sectors have a tremendous effect on climate change towards environmental crisis. Therefore, introducing battery electric-based mobility, specifically in public transportation because of the specific and predictable traveling schedule, provides a

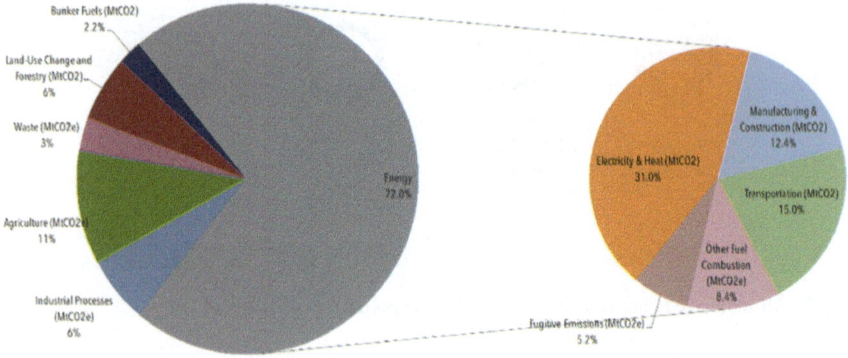

Fig. 3 2013 global manmade greenhouse gas emissions [9]

mutual contribution to both sectors. Whenever electricity is supplied from renewable resources to charging stations of proposed battery electric vehicles, the penetration level of renewable energy resources increases to supply energy for battery-based public buses. The power system can benefit when the buses are in off operation mode to support the grid via V2G options. Traveling schedules, routes, and traveling distances which are usually determined and predictable can ease the management and optimal operation of these technologies to reduce the adverse effects and improve the reliability and efficiency of related parameters in both power systems and transportation sectors. Figure 3 shows global sources of greenhouse gas emissions of which electricity, heat, and transportation have the highest share among all.

5 Findings of the Current Research and Extension to Real Case Study

The extension of this research will further investigate a case study in one of the developing countries such as Afghanistan. The public transportation in Afghanistan is highly affected politically and with poor management strategies. Large cities in Afghanistan, unlike other cities in the world, are following unreliable and deprived of technology public transport systems. Moreover, Afghanistan is a country in the state of infrastructure development or building stage. Therefore, the development of new technologies in the field of the power system and transportation, considering the optimal economic, technical, and environmental criteria, counts as one of the constructive affairs for the country. Scientific research in these fields can also be considered as effective steps towards the above goals.

As demonstrated in Fig. 3, the transportation sector holds the second position globally as the primary source of greenhouse gas emissions after electricity and heat. On the other side, Fig. 4 shows a fifteen years record of increasing vehicles in

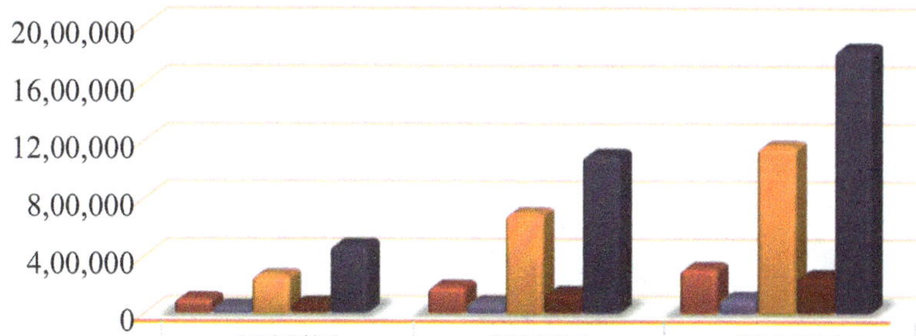

	2005/06	2010/11	2015/16
Lorries	1,00,883	1,84,799	3,11,905
Busses	41,731	74,834	1,04,543
Passenger Cars	2,62,700	6,91,573	11,56,215
Motorcycles	64,817	1,41,833	2,59,237
Total	4,70,131	10,93,039	18,31,900

Fig. 4 Number of vehicles and types in Afghanistan for fifteen years [10]

the country's transport sector which declares a serious issue in future management and impacts on climate change. Based on this data, the vehicles increased 290% from 2005 to 2015 which indicates an annual growth rate of 29% amazingly.

6 Opportunities and Challenges

Afghanistan has been involved in three decades of war which caused not only huge destruction of the country's infrastructure but paused the development of country almost in all aspects, specifically, technological, environmental, public health, economic growth, and social advancement as some major instances. Today Afghanistan is in the stage of building infrastructure and investment. Afghanistan is seeking to build the infrastructure so that it can effectively improve the current situation by considering future needs (Fig. 5). Less investment in the transportation sector is one of the challenges that require serious attention. Further investment of government and offering proper incentives as well as inviting private sectors within the necessary measures in place, such as proper mechanisms for monitoring, developing a secure investment environment, and implementing the appropriate policies opens many job opportunities besides many other advantages.

It is worthy of mentioning that the recent update of the country's transport sector master plan update (2017–2036) [11] has not included the concern of electric mobility, which shows a severe gap in the country's development in electric road

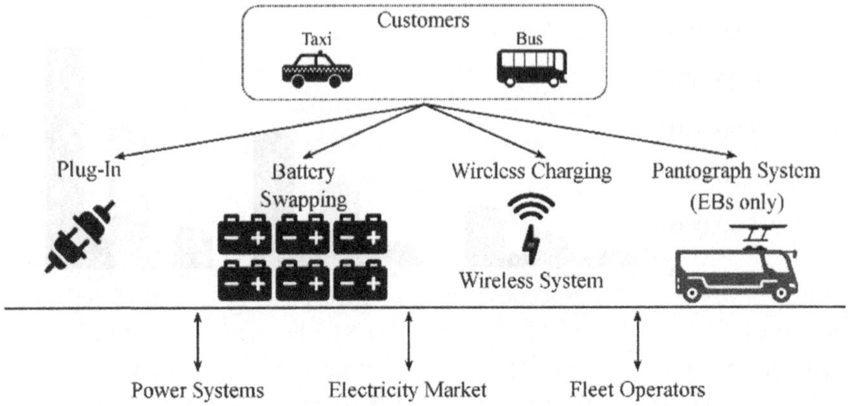

Fig. 5 Merging point of the electric power system and public transportation [2]

mass transportation. From another angle, Afghanistan is known to have the ability to exceed Bolivia as the first world producer of lithium. The two primary sources of lithium are hard rock sources in pegmatites and a solution within continental brines, both of which are said to be present in Afghanistan [12].

7 Electric Bus and Charging Stations

Electric bus refers to a public transport vehicle that can directly connect to electricity via cables as a pantograph system which was illustrated in Fig. 6 or to store energy on an on-board embedded battery storage system. The battery part of an electric bus is considered the most expensive part in which the world of electric bus producers is struggling in how to improve the energy density and decrease the associated cost. However, the lifetime cost or mean tone cost of the electric buses are the cheapest options due to fewer movement parts in their production comparing to nonelectric fuel-based buses.

The charging stations are another important facility which is operated to charge the vehicles. Charging of electric buses without control mechanism and management strategies significantly affects the electricity grid by changing the behavior of network performance including power loss improvement.

These infrastructures are usually divided into three categories: battery swapping station, plug-in charging station, and wireless charge stations [2]. Plug-in charging station as the most efficient and common type of charging station is referred to in this research. The word charging station in this paper is used to refer to the plug-in charging station unless otherwise defined.

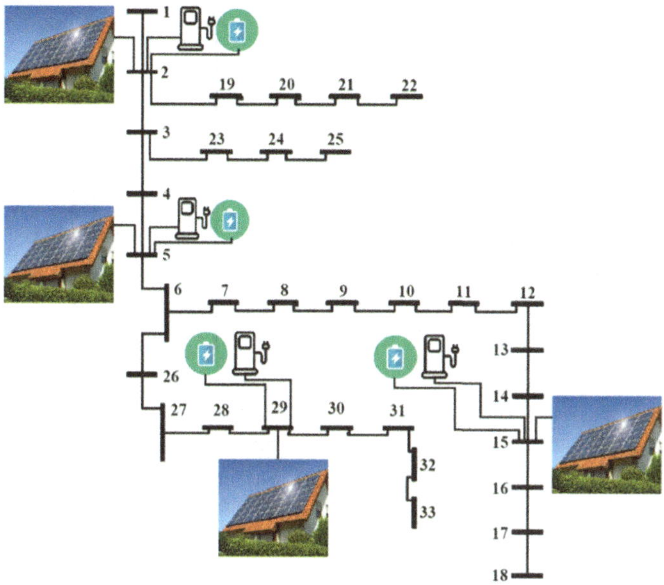

Fig. 6 Modified IEEE test distribution network

8 Proposed Model of Electric Bus Operation

The governmental locations are the most valid and feasible points in developing countries like Afghanistan to start this newly developed technology. The government employees usually have the scheduled sign in and sign out from their workplace with the huge numbers. The buses will have two long periods of opportunity to be charged at normal level two AC charging mode. This type of charging requires a longer time but lower power which reduces the adverse effect on the power grid, especially when the charging stations are supplied from the grid without any auxiliary renewable power generation.

To do so, in this study, four locations are identified as suitable and practically assumed available to introduce this technology. For each location, a charging station is assigned as indicated in Fig. 7. To alleviate not only the adverse effect of charging stations but also the network in a mutual contribution of the power system and transport sector, solar power and storage system are incorporated in each location associated with charging stations.

There are four locations considered based on assuming availability and as easier deployable governmental sites. Five electric buses for each location are considered to accommodate the transport of a maximum of 280 employees in each governmental organization. The round trip distance, the energy required, and the operation time of each of the buses are indicated in Tables 1, 2, 3 and 4.

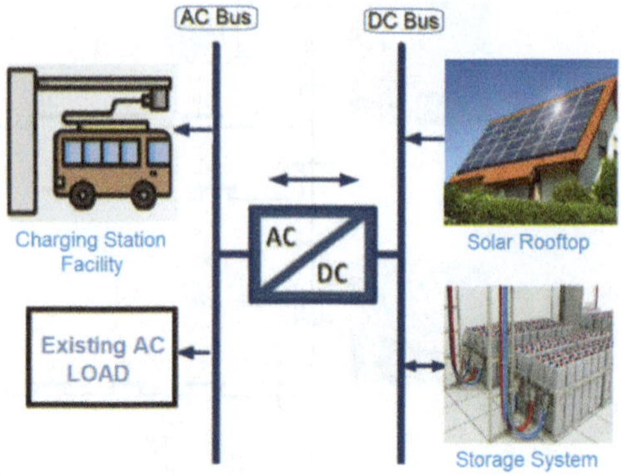

Fig. 7 Proposed schematic for candidate locations of E-bus charging station (in detail)

Table 1 Details of route, required energy, operation hours, and the number of stops related to location 1

Routes/electric bus	Round trip traveling distance (km)	Energy required (kWh)	Bus operation hours	Number of employees	Number of stops
R1/EB1	250 × 2	100 × 2	6:00–8:00 am/5:00–8:00 pm	50	10
R2/EB2	240 × 2	96 × 2	6:00–8:00 am/5:00–8:00 pm	45	7
R3/EB3	220 × 2	88 × 2	6:00–8:00 am/5:00–8:00 pm	48	8
R4/EB4	180 × 2	72 × 2	6:00–8:00 am/5:00–8:00 pm	46	5
R5/EB5	150 × 2	60 × 2	6:00–8:00 am/5:00–8:00 pm	40	7

Table 2 Details of route, required energy, operation hours, and the number of stops related to location 2

Routes/electric bus	Round trip traveling distance (km)	Energy required (kWh)	Bus operation hours	Number of employees	Number of stops
R1/EB1	280 × 2	112 × 2	6:00–8:00 am/5:00–8:00 pm	44	8
R2/EB2	200 × 2	80 × 2	6:00–8:00 am/5:00–8:00 pm	39	5
R3/EB3	170 × 2	68 × 2	6:00–8:00 am/5:00–8:00 pm	50	8
R4/EB4	180 × 2	72 × 2	6:00–8:00 am/5:00–8:00 pm	40	6
R5/EB5	230 × 2	92 × 2	6:00–8:00 am/5:00–8:00 pm	49	5

9 Methodology and the Test Case Study

In this paper, an IEEE test distribution network is considered to validate the technical concern on how to be managed using the optimization strategy. To step forward utilizing the recent technologies like electric mobility in transportation, incorporation

Table 3 Details of route, required energy, operation hours, and the number of stops related to location 3

Routes/electric bus	Round trip traveling distance (km)	Energy required (kWh)	Bus operation hours	Number of employees	Number of stops
R1/EB1	254 × 2	102 × 2	6:00–8:00 am/5:00–8:00 pm	48	6
R2/EB2	210 × 2	84 × 2	6:00–8:00 am/5:00–8:00 pm	47	4
R3/EB3	250 × 2	100 × 2	6:00–8:00 am/5:00–8:00 pm	42	7
R4/EB4	180 × 2	72 × 2	6:00–8:00 am/5:00–8:00 pm	46	5
R5/EB5	220 × 2	88 × 2	6:00–8:00 am/5:00–8:00 pm	50	5

Table 4 Details of route, required energy, operation hours, and the number of stops related to location 4

Routes/electric bus	Round trip traveling distance (km)	Energy required (kWh)	Bus operation hours	Number of employees	Number of stops
R1/EB1	242 × 2	97 × 2	6:00–8:00 am/5:00–8:00 pm	50	10
R2/EB2	220 × 2	88 × 2	6:00–8:00 am/5:00–8:00 pm	42	7
R3/EB3	180 × 2	72 × 2	6:00–8:00 am/5:00–8:00 pm	50	8
R4/EB4	230 × 2	92 × 2	6:00–8:00 am/5:00–8:00 pm	49	5
R5/EB5	190 × 2	76 × 2	6:00–8:00 am/5:00–8:00 pm	47	7

of renewable-based supply charging stations, energy storage systems, electric buses, and EVs into governmental sites of the proposed network is the interest of this paper to be investigated further. A multiobjective optimization approach is put forward to size the demanded solar PV, optimal battery storage capacity, its charging schedule, and optimal operation of charging stations. The solar data in [13] is utilized for this case study.

The objective functions are introduced as total daily power loss minimization and squeezing the charging/discharging power of the storage system which affects the size, battery degradation, and cost.

Equations 1 and 2 show the first and second objective functions mathematically.

$$F1 = \sum_{I=1}^{N_{br}} \sum_{h=1}^{h_{maz}} R_I \times I_{Ih}^2 \tag{1}$$

$$F2 = \sum_{h=1}^{h_{max}} \left| P_{ch}^h \right| + \sum_{h=1}^{h_{max}} \left| P_{disch}^h \right| \tag{2}$$

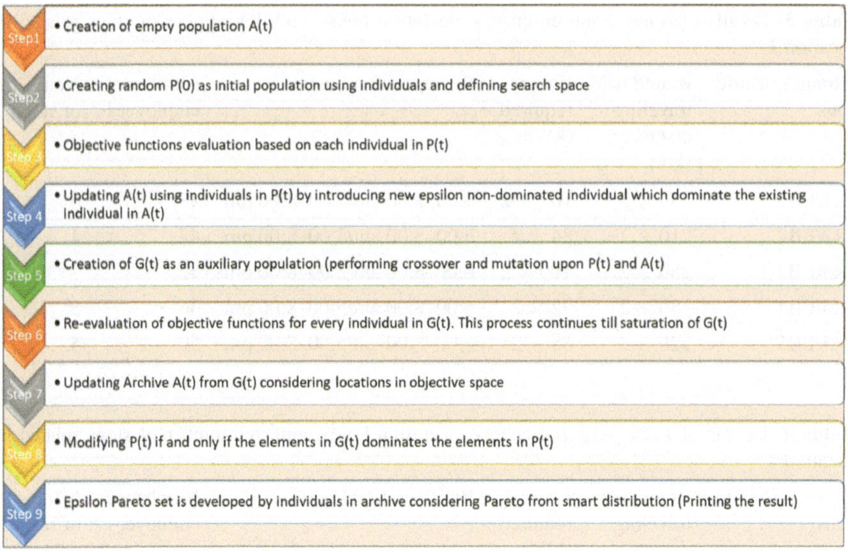

Fig. 8 Steps of the proposed optimization algorithm

where $F1$, $F2$, l, Nbr, h, P_{ch}^{h}, and P_{disch}^{h} are first, second objective function, branch number, the total number of branch numbers, hour, charging power, and discharging power of storage system, respectively.

A set of constraints is introduced to the load flow-based optimization formulation to bound the lower and upper limits for design variables as well as determined parameters. These constraints are power balance constraints, voltage deviation limit, maximum number of solar panels, battery constraints considering battery self-discharge rate and charge/discharge efficiency, PV inverter reactive power constraint, and transmission line thermal capacity constraints [14–16].

For conducting the proposed optimization problem, an epsilon multiobjective genetic algorithm is utilized which is further explored and exploited in [17]. The algorithm uses the epsilon dominance concept by utilizing three forms of main population P(t), archive population A(t), and auxiliary population G(t). A(t) stores the dominated solutions and is updated in each iteration. G(t) is obtained as an auxiliary population P(t) and A(t) by performing the two famous mutation and crossover operators over them. Figure 8 expresses the nine main steps of the algorithm process in summary.

10 Results and Discussion

A power flow approach using forward–backward sweep, in conjunction with the optimization algorithm, attempts to minimize the objective functions by establishing the set of determined constraints (Fig. 9). The network performance is an important concern when the new technologies are introduced to the distribution network

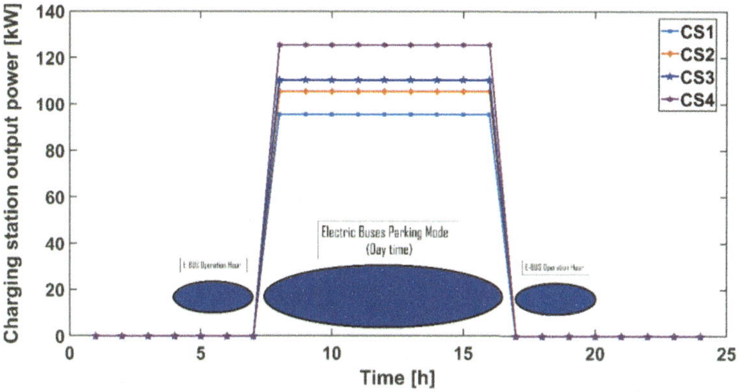

Fig. 9 Charging stations and electric buses operating in on and off mode

(Fig. 10). Without considering the effective methods of controlling and optimizing the new technologies introduced, effective grounding in the use of these elements, especially in power grids, presents serious challenges.

After exploiting the optimal point from obtained Pareto based on predetermined criteria and objective functions' priority, the results from the perspective of technical parameters in the test power distribution network are plotted and analyzed in this

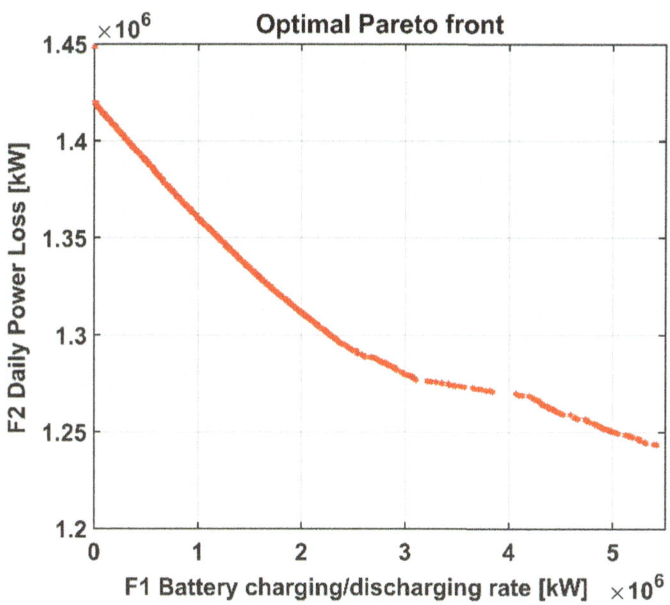

Fig. 10 Optimal Pareto front (results of epsilon multiobjective genetic algorithm)

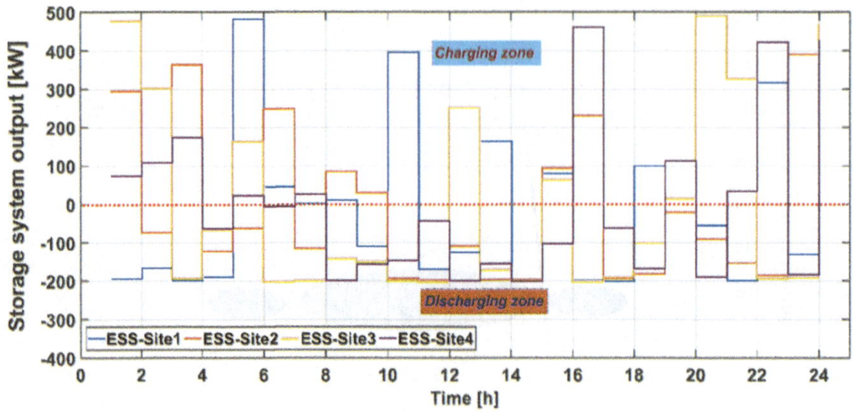

Fig. 11 Optimal charge/discharge scheduling of ESS in all candidate sites

section. Figure 11 shows the optimal power charging/discharging from each storage system at every candidate location.

The positive axis in the upper part of the red dotted line indicates a charging state from the battery to grid in order to compensate the electric bus's charging by storing extra power supply by solar power and during off-peak hours. Based on charging strategies assigned to the batteries and control mechanism of each battery state of charge as well as 60% depth of discharge (DoD), total charge and discharge to and from the battery are constrained to be almost equal.

Voltage rise and fall are one of the main issues when the charging stations and intermittent renewable energy resources integrate to the main grid. In this study, to maintain the network performance in good condition and withstanding an acceptable range of operation as well as power loss issues, the energy storage system is incorporated as discussed above. According to results and following this approach, not only voltage rise and fall is controlled, but also the total network power loss decreased from 1.5 to 1.4 MW on a daily basis compared to the base case.

Furthermore, the minimum voltage at node-18 improved from 0.9183 pu in the base case to 0.9255 pu in the optimal case after performing optimization, as shown in Figs. 12 and 13, respectively. Lastly, the mean voltage is also demonstrated for each case, which shows an improvement from 0.953 pu in the base case to 0.954 in the optimal case where solar power, storage system, and charging stations are incorporated.

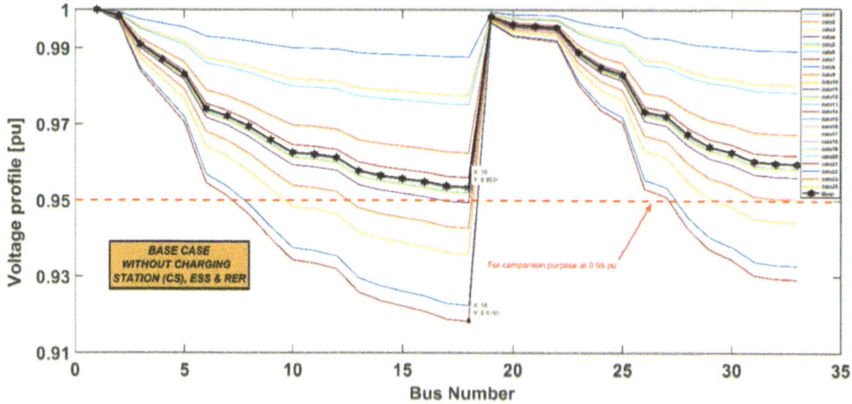

Fig. 12 Voltage profile for each node and every hour in the base case

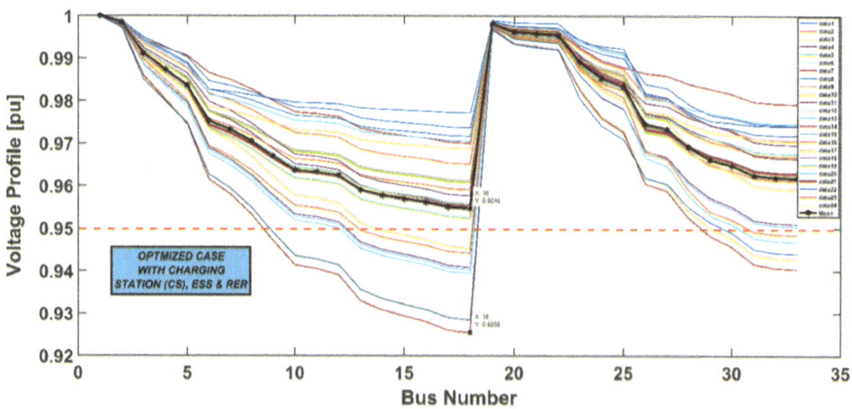

Fig. 13 Voltage profile for each node and every hour after optimization (optimal case)

11 Conclusion

An interesting topic that attracts recent research studies and researchers in the field of the power system and transportation through electric transportation technologies is proposed in this paper. Battery electric bus as a public transport facility and having predetermined traveling schedule, route, and distance was focused in this research to contribute mutually solving the raised environmental issues by the transportation sector as well as technical support to the grid and finally reliable mobility with great ride comfort. Meeting the related SDGs is discussed which is a global concern for developing and developed nations. A timely based program is simulated in MATLAB utilizing one of the powerful optimization tools to demonstrate the vital role of optimization in this topic as well as technical behavior of power distribution network when electric buses are incorporated. Specific governmental sites are selected, and optimal size and scheduling of solar power and storage system are attained successfully to supply the electric bus charging station, controlling their operation and proposed network parameters, simultaneously. The multiobjective approach succeeded in optimal Pareto front and the results exploited from a decision Pareto point validate the superiority of the proposed approach. Further economic and environmental analysis, enabling E2G option, electric bus to EV charging at nighttime as a novel approach as well as its deployment in a real case study are the extended future work of this paper.

References

1. Lin, Y., Zhang, K., Shen, Z.J.M., Ye, B., Miao, L.: Multistage large-scale charging station planning for electric buses considering transportation network and power grid. Transp. Res. Part C: Emerg. Technol. **107**, 423–443 (2019). https://doi.org/10.1016/j.trc.2019.08.009
2. Clairand, J.M., Guerra-Terán, P., Serrano-Guerrero, X., González-Rodríguez, M., Escrivá-Escrivá, G.: Electric vehicles for public transportation in power systems: a review of methodologies. Energies **12**, 3114 (2019). https://doi.org/10.3390/en12163114
3. Lin, Y., Zhang, K., Shen, Z.J.M., Miao, L.: Charging Network planning for electric bus cities: a case study of Shenzhen, China. Sustainability **11**, 4713 (2019). https://doi.org/10.3390/su11174713
4. Cheng, S., Li, Z.: Multi-objective network reconfiguration considering V2G of electric vehicles in distribution system with renewable energy. Energy Procedia **158**, 278–283 (2019). https://doi.org/10.1016/j.egypro.2019.01.089
5. Islam, M.R., Lu, H., Hossain, M.J., Li, L.: Mitigating unbalance using distributed network reconfiguration techniques in distributed power generation grids with services for electric vehicles: a review. J. Clean. Prod. **239**, 117932 (2019). https://doi.org/10.1016/j.jclepro.2019.117932
6. Shi, R., Li, S., Zhang, P., Lee, K.Y.: Integration of renewable energy sources and electric vehicles in V2G network with adjustable robust optimization. Renew. Energy **153**, 1067–1080 (2020). https://doi.org/10.1016/j.renene.2020.02.027
7. Iacobucci, R., McLellan, B., Tezuka, T.: Costs and carbon emissions of shared autonomous electric vehicles in a virtual power plant and microgrid with renewable energy. Energy Procedia **156**, 401–405 (2019). https://doi.org/10.1016/j.egypro.2018.11.104
8. United Nations (UN): Sustainable Development Goals (SDGs). https://sustainabledevelopment.un.org/sdgs. Last accessed 2019/11/01
9. Global Emissions. https://www.c2es.org/content/international-emissions/. Last accessed 2019/11/01
10. Second national communication under the United Nations Framework Convention on Climate Change (UNFCCC). National Environmental Protection Agency (NEPA) - Islamic Republic of Afghanistan, Kabul, Afghanistan (2017)
11. Afghanistan Transport Sector Master Plan Update (2017–2036). Asian Development Bank, Manila, Philippines (2017)
12. Schouker, P.: Afghanistan Has Only One Hope: Lithium. https://nationalinterest.org/feature/afghanistan-has-only-one-hope-lithium-18372. Last accessed 2019/11/01
13. Ahmadi, M., Lotfy, M.E., Shigenobu, R., Yona, A., Senjyu, T.: Optimal sizing and placement of rooftop solar photovoltaic at Kabul city real distribution network. Trans. Distrib. IET Gener. **12**, 303–309 (2018). https://doi.org/10.1049/iet-gtd.2017.0687

14. Ahmadi, M., Lotfy, M.E., Howlader, A.M., Yona, A., Senjyu, T.: Centralised multi-objective integration of wind farm and battery energy storage system in real-distribution network considering environmental, technical and economic perspective. Trans. Distrib. IET Gener. **13**, 5207–5217 (2019). https://doi.org/10.1049/iet-gtd.2018.6749
15. Ahmadi, M., Lotfy, M.E., Shigenobu, R., Howlader, A.M., Senjyu, T.: Optimal sizing of multiple renewable energy resources and PV inverter reactive power control encompassing environmental, technical, and economic issues. IEEE Syst. J. (2019)
16. Ahmadi, M., Yona, A., Senjyu, T.: Optimum consideration of electric vehicle/electric bus into distribution network incorporating ESS and maintaining sustainable transportation. In: Presented at the The IEEJ Technical Meeting, PE-2019, Okinawa, Japan (2019)
17. Ahmadi, M., Danish, M.S.S., Lotfy, M.E., Yona, A., Hong, Y.Y., Senjyu, T.: Multi-objective time-variant optimum automatic and fixed type of capacitor bank allocation considering minimization of switching steps. AIMS Energy **7**, 792 (2019). https://doi.org/10.3934/energy.2019.6.792

Smart and Sustainable Township: An Overview

Mozhdah Hafizyar, Ahmad Rasa Arsallan, Najib Rahman Sabory, Mir Sayed Shah Danish, and Tomonobu Senjyu

1 Introduction

Since the beginning of the twenty-first century, rapid urban population growth has challenged regional authorities and city inhabitants. Roughly 67% of the world's population lives in urban areas, and they are responsible for around 70% of the world's greenhouse gas emissions [1]. Rapid urban population growth strains city infrastructure and services like water, energy, transport, and other utilities. New urban planning models that take advantage of digital technology can be applied based on a city's resources, and smart cities rely upon a city's central infrastructure while intelligent cities integrate transportation systems, schools, businesses, open spaces, and government resources. The automation and smartness of cities as a microgrid (small-scale intelligent energy infrastructure) in various configurations (interconnected, radial, and hybrid) has attracted since 1995. Smart cities also consider financial and environmental impact on improving quality of life and security concerns.

From smart urban planning and modern architecture perspectives, the application of intelligent technologies and the internet of things (IoT) enables communities by using two-way interaction of a comfortable lifestyle [2]. Besides of IoT application in the context of urban planning and development, renewable energy technologies in buildings and urban mitigates carbon-intensive energy sources and greenhouse gas emission [3]. Lifestyle changes and sustainable housing are indebted to balanced urban planning in terms of sustainability pillars (technical, environmental, social, institutional, economic) integrated-application [4]. According to the reports, urban

M. Hafizyar (✉) · A. R. Arsallan · N. R. Sabory · M. S. S. Danish
Kabul University, Kabul 1006, Afghanistan
e-mail: mozhdah.hafizyar@outlook.com

M. S. S. Danish · T. Senjyu
University of the Ryukyus, Okinawa 9030213, Japan

M. S. S. Danish et al. (eds.), *Sustainability Outreach in Developing Countries*,
https://doi.org/10.1007/978-981-15-7179-4_5

areas consume around 70% of energy, with a significant share of 40% is recorded for building energy consumptions [5].

Affordable energy access contributes towards environmental health, boost socio-economic (create job opportunities, improve household incomes, increase health and education quality, and enhance lifestyle). Afghanistan with about 31 million populations, around 30% in urban and 10% in rural areas, have access to electricity [6, 7]. Besides the low rate of access to clean energy, heavy reliance on energy import, deforestation, and health issues due to using primary energy resources have remained significant challenges in Afghanistan [8]. Primary energy sources (nonrenewable energy—fossil fuels: coal, crude oil, natural gas, nuclear fuel) captures directly from the environment. Whereas, the secondary form of energy sources (renewable energy: hydropower, biomass, solar, wind, geothermal, and ocean energy) produces or converts from the primary energy sources [2]. Mostly the Kabul city loads are residential, in which residential buildings consume an estimated one-third of total primary energy resources [9]. A city with a high density of population and lack of access to clean and stable energy will lead to critical environmental challenges. Therefore, preventive and corrective solution mechanisms can mitigate the dilemma. As part of these efforts, this study deals with an exhaustive survey of the subject.

Overall, this study enumerates the primary potential and available opportunities for smart urban planning and design in the context of least developing countries. However, fulfilling these efforts seems complicated in the form of sustainability criteria requirements. This study is embraced a case study of Kabul city with the format and informal settlements, along with a general overview of existing gaps in terms of urban planning and design, as well as tried to explore emerging solutions as a future direction for students, researchers, urban planners, and interested stakeholders. In Kabul, the unplanned, informal settlement of Bibi Mahro is adjacent to the 3rd Macrorayan area, which is a planned and formal settlement urban segregation that exists between these areas so socio-cultural, spatial, physical, and economic differences should be addressed in a successful integration plan. The smart and sustainable township concept in this paper focuses on designing an eco-friendly environment, smart and sustainable societies, and urban security using smart city criteria.

2 Kabul Urban Design Framework (KUDF) Contribution to the Sustainable Development Goals (SDGs) Deployment

Kabul the capital of Afghanistan, located in the eastern section of the country with more than 4 million inhabitants. The KUDF has been constructed on the basis of sustainable development principles, and its implementation will help Kabul meet its residents' needs and provide a suitable environment for future generations. The key challenges addressed by the KUDF include challenges in housing, economic infrastructure, regional and urban mobility, public space and recreation, water management, and equitable access.

KUDF's strategy for the Massoud Boulevard area of Bibi Mahro, "Creating an ecosystem of innovation and economic development" [10] is consistent with SDGs of quality education by developing an education cluster that includes hospitals and businesses to "ensure inclusive and equitable quality education and promote lifelong learning opportunities for all" [11]. A strong education cluster will create opportunities for "equitable social and economic regeneration" [10]. Consistent with the SDGs for life on land, a network of public, open spaces that preserve native vegetation and trees with social anchors like women centers, mosques, and schools should promote biodiversity and a better quality of life. Native greenery and an extensive transit network control the level of air pollution. Specific suggestions for Massoud Boulevard in the KUDF points to a vision of an urban ecosystem that meets the SDGs and includes specific suggestions for transit stops, office complexes, plazas, housing, community centers, local parks, and commercial/industrial development.

2.1 Urban Segregation and Social Exclusion Challenges

Kabul faces challenges in addressing poverty and social inequality in its informal settlements. Contrasting environments between the informal settlement in Bibi Mahro and the formal settlement in 3rd Macrorayan has created urban segregation. As it defined: "the non-random allocation of people who belong to different groups into social positions and the associated social and physical distances between groups" [12]. People in the upper social class generally live in rich and upscale Macrorayan. At the same time, Bibi Mahro—valued as a historic site has a large concentration of poor people, considered as a low-prestige area. The stigma, segregation, and social exclusion of those living in Bibi Mahro reflect processes related to intractable cycles of urban poverty (e.g., unemployment, low income, illiteracy). The Kabul airport road is an established boundary that separates these two settlements that further highlights the social segregation and lack of communication as well as social-link between them.

Resolving urban segregation between Bibi Mahro and Macrorayan areas is a priority in developing a sustainable, smart township plan. Deliberate strategies that integrate and improve the economic, physical, social and environmental conditions of an area within the city is imperative.

2.2 Solution Mechanism

Good design encourages sustainable use of shared resources. Properly planning, denser cities should reduce overexploitation of natural resources and facilitates everyday living by enabling equal access to land, food, and water. In designing this township, authors have focused on water management, energy, transportation, housing and technology, and have integrated them in these plans to identify, revitalize, protect, and produce high-quality public spaces (Figs. 1 and 2). The design strategy has two parts: (1) Sustainability, (2) Smart features. A smart and sustainable city's goals are adaptable, reliable, scalable, accessible, and resilient:

- Improve quality of life;
- Ensure economic growth with better employment opportunities;

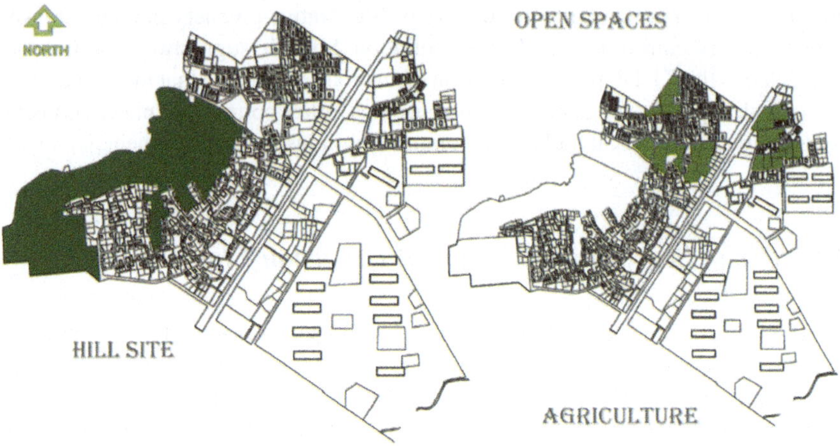

Fig. 1 Agricultural and open space in site

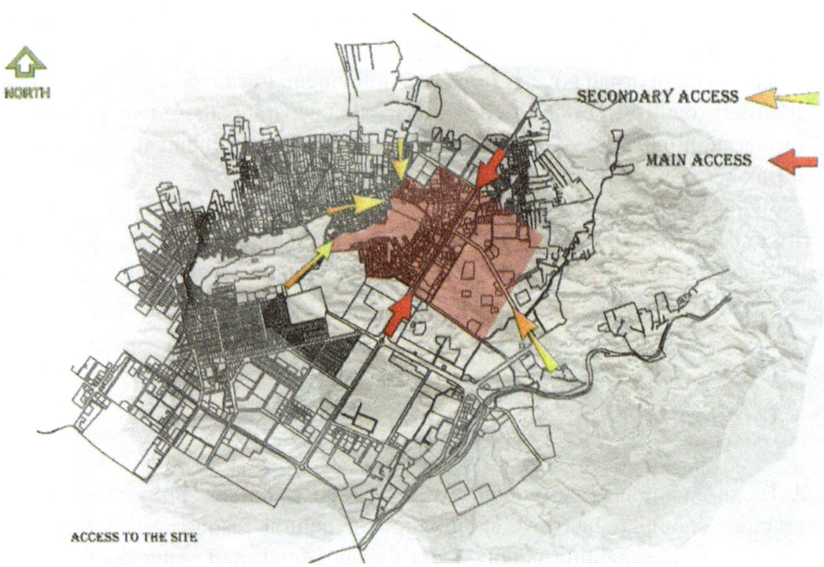

Fig. 2 Site map and access routes

- Improve citizens' well-being by ensuring access to social and community services;
- Establish an environmentally responsible and sustainable approach to development;
- Ensure efficient delivery of basic services and infrastructure, such as public transportation, water supply and drainage, telecommunication and other utilities;
- Address climate change and environmental issues;
- Provide an effective regulatory and local governance mechanisms that enforce equitable policies [13].

3 Methodology

Definitions for the term smart city and sustainability were compared in a systematic review of the literature focused on integrating formal and informal settlements. These concepts were applied in our case study of designing the Bibi Mahro/3rd Macroyan site. Our goals were to enhance the quality of life and provide a sustainable environment. We first integrated the formal and informal settlements to provide and improve the area's economic, social, physical, and environmental conditions.

Designing hillside areas should address significant issues like street design, grading, parking, drainage, sewer availability, architecture, visual landscaping impacts, preservation of natural features, fire access defensibility, and geologic hazards. Surveys of the site revealed the importance of the following considerations:

- Predominant building materials are stone, brick, concrete, steel, and clay.
- Most of the buildings did not consider the local climate in their designs.
- 25% of the site is classified as a service area, 20% is vacant, 15% retail business, and the remaining residential.
- Vacant lots designated for parking were absent. Lacks of parks or recreational green space.

This case study provides a detailed plan to make the site more sustainable and smart, with priorities on energy (production and conservation), water and wastewater management, transportation, and utilization of intelligent technology.

4 Sustainable Solutions

4.1 Access to Affordable Energy

As said, buildings consume about 40% of total energy in urban areas with a high share of 30% greenhouse gases (GHGs) emissions [14, 15]. Encouraging energy conservation within buildings to create an energy-efficient environment promotes sustainability. Afghanistan is an importing energy country that imports around 80%

of its needs from the neighboring countries [16]. Currently, about 30% of households in Afghanistan connected with noncontinuous power supply (approximately 117.2 kWh/year per capita), ranges from zero in rural to near 100% electricity supply in cities [17].

One electric provider serves the citizens of Afghanistan: Da Afghanistan Breshna Sherkat (DABS). While it produces energy from hydroelectric dams, it must procure power from other neighboring countries to meet demand from citizens and factories for heating, cooling, and ventilation. This demand stresses the DABS grid and causes short-term outages. While the initial costs may be high, investing in alternative sources for energy should relieve much of this demand at a lower cost in the long run. New buildings can be built with Thermal Solar Radiators, which capture solar energy to maintain building temperature and provide an additional source of energy. Solar panel calculations are given in Table 1 (Fig. 3), along with the following specifications:

- Residential Load demand = 1.5 kW
- Commercial Load demand = 0.7 kW/0.5 kW
- 1 W solar panel cost = \$6

Table 1 Solar panel calculations for one building load demand

No.	Land use	1 Story	3 Story	Total (6 Story)	Total area	Total cost
1	Residential	18 panel	54 panel	120 panel	129.6 m^2	180,000 \$
2	Commercial	22 panel	66 panel			

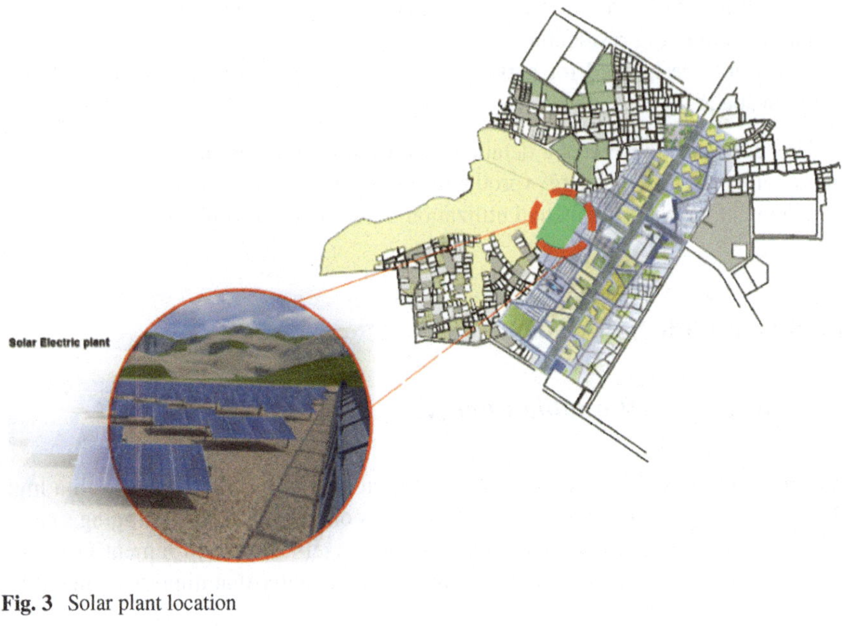

Fig. 3 Solar plant location

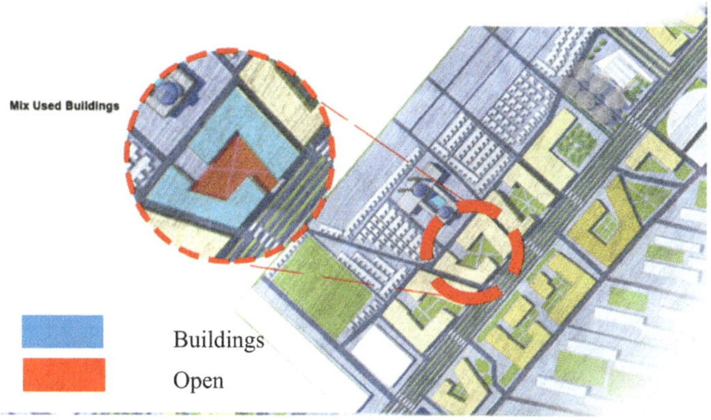

Fig. 4 Building layout

- One panel load = 250 W
- Area of one panel = 1.08 m^2
- Area of one mix use building is 11,318 m^2
- 200 amp battery for electricity storage
- Inverter to convert AC to DC
- Roof angle estimated 45°.

To maximize these benefits, we propose that DABS purchase an additional 120 panels for another solar power plant that serves the township. The extra energy produced can be used for transportation, commercial use, and residents who are unable to afford a rooftop panel.

4.2 *Social Engagement*

A neighborhood plan (Fig. 4) that encourages productivity and social interaction contribute to a sustainable environment.

Our planned neighborhood consists of two parts, as illustrated in Fig. 5.

- Mixed-use buildings.
- Open green spaces, divided for recreation and urban farming.

4.3 *Water Management*

Capturing and recycling wastewater is critical for sustainability. Rainwater overflow in Kabul has been a historical problem that inconveniences residents by flooding streets and sewers, hindering public movement and transport, damaging or destroying

Fig. 5 Farming and recreational area

mud houses built by the poor. These problems are especially unsatisfactory during rainy seasons (Fig. 6), and a centralized system of effective water management is needed.

Farming will create additional economic opportunities while promoting energy conservation. The following are strategies for collecting and utilizing rainwater in this township for irrigation.

- Strategy 1: By storing rain runoff in tanks below streets, 0.4 m^3 water can be saved per meter of road per year. This water can be used simply for irrigating trees, farms, parks, and other green spaces.
- Strategy 2: Storing rainwater from building gutters for farming. Water stored in these tanks will be filtered by layers of sand, gravel and pebbles to remove contaminants and provide clean water for farming. A drip irrigation system further raises the efficiency of water usage for farming.
- Strategy 3: Canopies can collect water in marketplaces and be stored in the tanks underneath the canopies. These canopies will also provide shade and comfortable ventilation for the people visiting the market.
- Strategy 4: Parks and sidewalks will be designed to replenish underground water supplies for ponds and rain gardens to ensure that the water table is at the correct desired level. Building underground street tunnels can have long-term benefits for energy management and municipal maintenance costs.
- Strategy 5: A network of underground street tunnels can be a cost-efficient energy-saving strategy. Each tunnel or track can be specialized and repurposed as necessary, saving time and expense for maintenance and re-excavating for new projects in the future.

Fig. 6 Rainy season and condition of some streets in Kabul city

On side streets, planter strips (5–112 feet wide) should be placed between curbs and sidewalks to allow for landscaping and a greater separation between pedestrians and autos unless excessive grading is needed. These strips will absorb street water and hazardous gas emissions while providing fresh air. Sidewalks can also be built with materials that absorb and filter rainwater.

4.4 Transportation Facilities

Air pollution from daily traffic activities in the absence of an effective public transport system play a large role in energy consumption. Making this sector more sustainable would conserve energy, reduce hazardous emissions, and strengthen economic

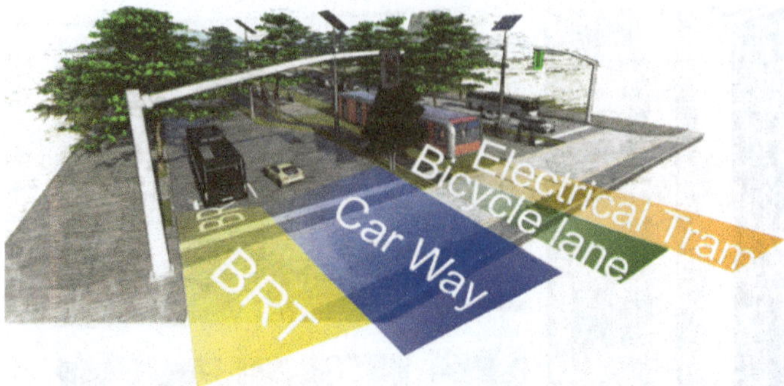

Fig. 7 The proposed transport illustration

development. In this plan, transportation and traffic challenges in this township are addressed with separate lanes (Fig. 7) for:

- BRT
- Express electric urban trams/buses
- Bicycles.

Fueled by electricity from the solar electric plant, electric trams provide express transportation between two designated major hub stations along the main street. BRT buses also travel along the main avenue but stop at short-range stations. A designated, separate bicycle lane should encourage citizens for a healthy transportation alternative that can reduce street noise and lower emissions. Our design places the bicycle lane next to the express tram lane to give a choice to bicyclists to park their bikes at a hub and ride the express tram or BRT. Real-time status for the trams and buses (location, estimated arrival time) will further inform and incentivize citizens.

4.5 Socio-economic Development

In this plan, we rely on innovative technology to develop a modern and environmentally-friendly urban market. Canopy "trees" consist of internal smart glass layers that cover a high-performing wooden skeleton and an external waterproof glass layer with building-integrated photovoltaic cells (BIPV). The BIPV glass stores enough power to support phone charging stations and to light the entire urban market during the evening using LED's incorporated into the wooden skeleton. The smart-glass layer can change its opacity in response to the intensity of daylight, controlling the temperature of the space underneath the canopy while allowing enough light underneath; during more cloudy days and at nighttime, the smart-glass would be transparent. The overhanging branches that feed into the trunks serve to collect rainwater to underground reservoirs beneath the plaza pavement. The collected water can be used for cleaning the market area. Water vaporization from the reservoirs also cools any breezes in the open-air market to further regulate the temperature during hot days. Hot or polluted air can escape through gaps in the canopy's "tree crowns" [18]. This dynamic construction reminds residents and visitors of the harmony of designing spaces that integrate ecological changes with architecture and society. Indeed this unique gathering space may become an iconic cultural attraction that anchors further development.

4.6 *Waste Management*

Waste management improves the psychological well-being of the neighborhood and can be leveraged for energy production. Public awareness and education are necessary to make progress in this plan. Environmental challenges, poor air quality, and density of the population ranked Kabul the 7th pollutant city in the world, as shown in Table 2 [19].

In the aftermath of recent civil wars, Kabul has suffered from a lack of central sewage and waste management system and a high rate of illiteracy. As a result, garbage and waste are scattered in many alleys and streets (Fig. 8). We propose the following strategies for effective and organized waste management in this township.

Table 2 The most top ten pollutant cities in the world

Rank	City	Pollution index	Exp. pollution index
1	Accra, Ghana	97.88	177.34
2	Tetovo, Macedonia	96.85	177.01
3	Faridabad, India	96.24	174.46
4	Ghaziabad, India	96.10	174.32
5	Kathmandu, Nepal	95.54	173.46
6	Ulaanbaatar, Mongolia	94.93	175.33
7	Kabul, Afghanistan	94.84	171.67
8	Cairo, Egypt	93.79	169.84
9	Dhaka, Bangladesh	92.96	167.41
10	Ho Chi Minh City, Vietnam	92.79	168.02

Fig. 8 Kabul unplanned areas alley

- Strategy 1: Different-colored recycles bins to separate recyclable and non-recyclable garbage throughout the town. Early separation by citizens will make separating these items more time- and energy-efficient.
- Strategy 2: Waste products can supply heat and electricity for households.
- Strategy 3: Smart waste containers throughout the city can maintain cleanliness while conserving energy. Sensors attached to the container inform the collection entities when the container is full and needs to be serviced. This smart technology can result in more efficient use of garbage trucks, saving gasoline, time, environmental impact, and money.
- Strategy 4: An effective recycling system using anaerobic baffled reactors (ABR method) can filter wastewater to create biogas for fuel, biomass fertilizer for agriculture, and filtered water for irrigation.

Some advantages and disadvantages of the ABR method are listed as following.

1. Advantage:
   - Resistant to organic and hydraulic shock loads
   - No electrical energy is required
   - Low operating costs
   - Long service life
   - Low sludge production as the sludge is stabilized

- Moderate area requirement (can be built underground).

2. Disadvantages:

   - Requires expert design and construction
   - Low reduction of pathogens and nutrients
   - Effluent and sludge require further treatment and/or appropriate discharge.
   - In this plan, this treatment facility is located outside the township.

4.7 Smartness and Automation

Smart technology brings the potential and promises to make many management processes more effective and efficient to improve the quality of life for residents. Smart features that support sustainability goals are included in the following examples:

- Smart switches to control solar PV energy management
- Smart monitoring water temperature in Thermal Solar Radiators to supply hot water and electricity to buildings
- Light-sensitive and motion-sensitive sensors to dim or control lights intensity (buildings and street lights).

Smart features that require connection with data in interconnected sensors and networks are considered in the following examples:

- Smart traffic signals (to control and regulate traffic during peak traffic)
- Smart community surveillance and security (including drones) for fire suppression, crime control, and security enhancement
- Smart parking management (parking availability notification)
- Smart air purifiers (monitor air quality continuously or each minute)
- Smart cybersecurity
- Smart dustbins and recycle bins
- Smart Wi-Fi stations and spots.

Smart technology can help conserve water and reduce waste in the township.

- Effective irrigation processes can rely on modular sensors that report real-time measurements of temperature, humidity, CO_2, rainfall, air quality, wind speed, light duration, soil moisture, soil temperature, pH, and nutrient composition.
- Sensors that monitor water conduits can reduce waste by locating pipes leakages.

5 Recommendations

Environmental problems in Afghanistan's cities, including Kabul, significantly contribute to health issues. In Kabul, air pollution levels can be extremely hazardous. Afghanistan's National Protection Agency reported the amount of particulate matter in the first 250 m from the ground is over three times higher (527 $\mu g/m^3$) than the recommended upper limit (150 $\mu g/m^3$) [20]. Here are some recommendations directed to Kabul Municipality, Ministry of Urban Development and Land, Ministry of Transport, Ministry of Agriculture, Irrigation and Livestock, Ministry of Energy and Water, and National Environment Protection Agency of Afghanistan:

- Use smart features to purify the air and maintain a clean environment.
- Use modular sensors for irrigation water management.
- Use security drones to help address security problems.
- Use ABR, underground reservoirs, rain recycling containers, ponds, and rain gardens to collect and recycle water for irrigation.
- Manage waste as an energy source for producing heat and electricity while creating employment opportunities.
- Use smart waste containers and light sensors to increase quality of life and improve services through efficient use of time, energy, and money.
- Create urban farming areas to spur economic growth and energy conservation cost-effectively. Farms and parks contribute to sustainable environmental use, aesthetic landscaping, and improved air quality.
- Build a solar power plant along the Bibi Mahro hillside on available property. Install rooftop solar panels and smart technology in mixed-use buildings in the settlement as the electricity generated can offset the installation costs. Excess power can be sold to DABS. This solar plant provides employment opportunities.

6 Conclusion

Urban planning and development recognize the importance of creating sustainable, smart, ecofriendly, and healthy cities in confronting critical problems of population growth, rapid urbanization, industrialization, and high usage of private vehicles. During the last few decades, smart city and sustainable development have become popular topics that moved from scholarly discussions among the fields of environmental economics, technology and science, urban planning, development, and management, to urban policymakers and professional practitioners. This paper presented the concepts of Smart city and sustainable township by designing a sample plan for Bibi Mahro and 3rd Macrorayan in Kabul. This plan encourages the development of an integrated urban area (Bibi Mahro and 3rd Macrorayan) that promotes smart and sustainable development and enhances physical, social, and environmental life. Incorporating smart technology and processes can be valuable in addressing urban management challenges in developing countries like Afghanistan.

References

1. Estevez, E., Lopes, N., Janowski, T.: Smart Sustainable Cities: Reconnaissance study. United Nations University Operating Unit on Policy-Driven Electronic Governance, Tokyo, Japan (2016)
2. Danish, M.S.S., Sabory, N.R., Ershad, A.M., Danish, S.M.S., Yona, A., Senjyu, T.: Sustainable architecture and urban planning through exploitation of renewable energy. Int. J. Sustain. Green Energy **6**, 1–7 (2016)
3. Danish, M.S.S., Matayoshi, H., Howlader, H.O.R., Chakraborty, S., Mandal, P., Senjyu, T.: Microgrid planning and design: resilience to sustainability. In: 2019 IEEE PES GTD Grand International Conference and Exposition Asia (GTD Asia), pp. 253–258. IEEE, Bangkok, Thailand (2019). https://doi.org/10.1109/GTDAsia.2019.8716010
4. Danish, M.S.S., Senjyu, T., Funabashia, T., Ahmadi, M., Ibrahimi, A.M., Ohta, R., Rashid Howlader, H.O., Zaheb, H., Sabory, N.R., Sediqi, M.M.: A sustainable microgrid: a sustainability and management-oriented approach. Energy Procedia **159**, 160–167 (2019). https://doi.org/10.1016/j.egypro.2018.12.045
5. Danish, M.S.S., Senjyu, T., Ibrahimi, A.M., Ahmadi, M., Howlader, A.M.: A managed framework for energy-efficient building. J. Build. Eng. **21**, 120–128 (2019). https://doi.org/10.1016/j.jobe.2018.10.013
6. Ibrahimi, A.M., Sediqi, M.M., Howlader, H.O.R., Danish, M.S.S., Chakraborty, S., Senjyu, T.: Generation expansion planning considering renewable energy integration and optimal unit commitment: a case study of Afghanistan. AIMS Energy **7**, 441–464 (2019). https://doi.org/10.3934/energy.2019.4.441
7. Ludin, G.A., Matayoshi, H., Danish, M.S.S., Yona, A., Senjyu, T.: Hybrid PV/wind/diesel based distributed generation for an off-grid rural village in Afghanistan. J. Energy Power Eng. **11** (2017). https://doi.org/10.17265/1934-8975/2017.02.003
8. Danish, M.S.S., Sabory, N.R., Danish, S.M.S., Senjyu, T., Ludin, G.A., Noorzad, A.S., Yona, A.: Electricity sector transitions in an after war country: a review of Afghanistan's electricity. J. Energy Power Eng. **11** (2017). https://doi.org/10.17265/1934-8975/2017.07.008
9. Danish, M.S.S., Senjyu, T., (eds.): Green building efficiency and sustainability indicators. In: Green Building Management and Smart Automation, pp. 128–145. IGI Global (2020). https://doi.org/10.4018/978-1-5225-9754-4
10. Sasaki: Kabul urban design framework. Ministry of Urban Development and Housing, Kabul, Afghanistan (2017)
11. United Nations (UN): Sustainable Development Goals (SDGs). https://sustainabledevelopment.un.org/sdgs. Last accessed 2019/11/01
12. Bruch, E.E. and Mare, R.D., 2008. Segregation processes. Appears in the Oxford Handbook of Analytic Sociology, edited by P. Bearman and P. Hedström
13. Trindade, E.P., Hinnig, M.P.F., da Costa, E.M., Marques, J.S., Bastos, R.C., Yigitcanlar, T.: Sustainable development of smart cities: a systematic review of the literature. J. Open Innov. Technol. Market, Complex. **3**, 11 (2017). https://doi.org/10.1186/s40852-017-0063-2
14. Bourdeau, M., Zhai, X. qiang, Nefzaoui, E., Guo, X., Chatellier, P.: Modeling and forecasting building energy consumption: a review of data-driven techniques. Sustain. Cities Soc. **48**, 101533 (2019). https://doi.org/10.1016/j.scs.2019.101533
15. Nematchoua, M.K., Yvon, A., Roy, S.E.J., Ralijaona, C.G., Mamiharijaona, R., Razafinjaka, J.N., Tefy, R.: A review on energy consumption in the residential and commercial buildings located in tropical regions of Indian Ocean: a case of Madagascar island. J. Energy Storag. **24**, 100748 (2019). https://doi.org/10.1016/j.est.2019.04.022
16. Danish, M.S.S., Senjyu, T., Sabory, N.R., Danish, S.M.S., Ludin, G.A., Noorzad, A.S., Yona, A.: Afghanistan's aspirations for energy independence: water resources and hydropower energy. Renew. Energy **113**, 1276–1287 (2017). https://doi.org/10.1016/j.renene.2017.06.090
17. Danish, M.S.S., Sabory, N.R., Danish, S.M.S., Senjyu, T., Ludin, G.A., Noorzad, A.S., Yona, A.: Electricity Sector Development Trends In An After-War Country: Afghanistan aspiration

for an independent energy country. J. Energy Power Eng. **11**, 553–557 (2017). https://doi.org/10.17265/1934-8975/2017.08.007

18. ArchDaily: Sustainable Market Square Competition Entry. https://www.archdaily.com/300920/sustainable-market-square-competition-entry-nikolovaaarso-na/. Last accessed 2020/04/01
19. Danish, M.S.S., Zaheb, H., Sabory, N.R., Karimy, H., Faiq, A.B., Fedayi, H., Senjyu, T.: The road ahead for municipal solid waste management in the 21st century: a novel-standardized simulated paradigm. In: The 3rd International Conference on Energy and Environmental Science 2019 (ICEES 2019). IOP Conference Series: Earth and Environmental Science (EES), Seoul, South Korea (2019)
20. Strategic Nahia Action Plan (SNAP): UN-Habitat, Kabul, Afghanistan (2018)

An Empirical Analysis of Sustainability Indicators in an Administrative Complex Design from Urban Planning Perspective

Hameed Shirzad, Ahmad Zia Amini, Yasser Qudir, Zakia Husssainy, Najib Rahman Sabory, Mir Sayed Shah Danish, and Tomonobu Senjyu

1 Introduction

Many studies are concerned with urban project planning and design aligned with sustainability requirements to ensure current needs and avoid negative impact on future generations. Cities with high density have a higher pollution rate due to an increasing population, lack of urban growth, and designing solutions with little consideration of sustainability goals. The globalization of knowledge sharing enables urban planners to ensure multidimensional factors considering historical features, territory development, innovative and smartness options, and existing urban planning regulations within sustainability requirements.

The SDGs 2030 Agenda declared in 2015, principally deals with 17-goals and 169-targets to enhance life quality (humanity), and combat climate change [1]. Sustainable development is a decision-making approach to overcome societies challenges and offer a balance of services within sustainability pillars [2]. However, sustainability requirements and its pillars are reported differently based on the scope and target area of different studies; in a general perspective, sustainability pillars are the balance of social, economic, technical, institutional, and environmental factors for long-run sustainability [3].

Apart from the sustainability concept, there are some essential factors to be adapted for an acceptable urban or building planning and design. These primary factors are pointed out in [4] as below:

- Health and hygiene
- Indoor air quality

H. Shirzad (✉) · A. Z. Amini · Y. Qudir · Z. Husssainy · N. R. Sabory · M. S. S. Danish
Kabul University, Kabul 1006, Afghanistan
e-mail: shirzad.arch@gmail.com

M. S. S. Danish · T. Senjyu
University of the Ryukyus, Okinawa 9030213, Japan

M. S. S. Danish et al. (eds.), *Sustainability Outreach in Developing Countries*,
https://doi.org/10.1007/978-981-15-7179-4_6

- Noise and acoustic
- Environment
- Safety
- Controllability
- Serviceability
- Adaptability
- Living quality
- Energy efficiency
- Use of renewable energy technologies
- Water conservation
- Automation
- Reuse options
- Waste management
- Cultural character
- Building economics
- Security
- Nature and heritage conservation.

Urban planning is stated interlink efforts that can be categorized into urban project analysis, urban regulation, and finally, urban planning (more details are given in [5]). This lack of sustainable planning manifests itself with poor transportation systems, lack of green spaces, and substandard building construction [6]. In recent years, Kabul city has quickly grown into one of the most populated cities in the world, but it has been unable to manage and support its growth. Poor construction, weak transportation systems, and building over potential green spaces negatively affect environmental and social sustainability [7]. Recently, the Government Administrative Complex (GAC) in Dar-ul Aman is proposed by the government of Afghanistan to mitigate city traffic and population density. This complex will approximately relocate all governmental offices from the city center to the southern edge of Kabul. Advantages to this solution including encouraging a central location for increased collaboration and providing an identity to that area. Disadvantages include accessibility problems created by traffic, pollution, demolition of agricultural or green spaces, lack of groundwater recharge, and special attention compared to other areas needing support.

2 Problem Statement

One of the most ambitious contemporary developments in Kabul is the GAC that affirms the historical symbolism of the governmental castle as a strong public symbol of Afghanistan rebirth after long years of wars and destruction. Around 100 ha are allocated to this project at the east of Dar-ul-Aman Palace near Deh-Dana Road (7th district). This project covers 30 governmental offices, which include 18 ministries, 10 governmental authorities, 2 banks, and other essential facilities. This site is estimated to take up one million m^2 with a 1.5 million m^2 building footprint. This development

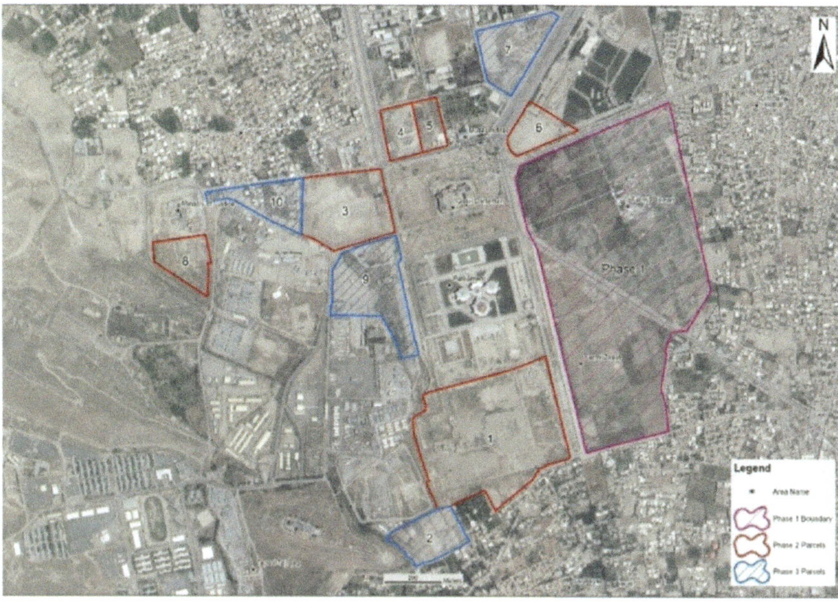

Fig. 1 The GAC project implementation phases [8]

moves major governmental activities from the city center to a suitable environment that provides better accessibility, security, and ease of operations. The relocation of ministries and governmental authorities will bring more effective coordination and communication, provide facilities for visitors, and develop the surrounding areas. Finally, this project can open employment opportunities to millions of unskilled, semi-skilled, and fully skilled job-seekers. This project will be implemented in three phases (Fig. 1).

- Phase 1 relocates the following governmental organizations: Ministry of Finance, Central Bank, National Procurement Authority, Pashtany Bank, National Bank, Ministry of Economy, Ministry of Mines and Petroleum, Capital Region Independent Development Authority, Afghanistan National Disaster Management Authority, Ministry of Urban Development and Housing, Ministry of Public Works, Ministry of Transport and Aviation, Independent Directorate of Local Governance, and Afghanistan Independent Land Authority.
- Phase 2 relocates these offices: Fuel and Liquid Interise, Ministry of Commerce and Industry, Supreme Court, General Attorney Office, Ministry of State for Parliamentary Affairs, Ministry of Religious Affairs and Haj, Ministry of Women Affairs, Ministry of Borders and Tribal Affairs, Ministry of Refugees and Repatriation, General Independent Directorate of Kochi Affairs, Ministry of Labor, Social Affairs, Martyrs and Disabled, Ministry of Public Health, Ministry of Counter Narcotics, Ministry of Information and Culture, Ministry of Education, and Writers Association Union.

Fig. 2 The GAC master plan [8]

Developed with a master plan (Fig. 2) focused on infrastructure, building design, and review services, the GAC is conceived according to consensus principles from the project steering committee:

- Mixed-use governmental district: Promote a 24 h secure and lively mixed-use neighborhood where people live, work, and play.
- Integration with surroundings: Preserve the existing urban fabric but integrate with additional uses (educational, administrative, commercial, governmental, residential, and planned areas).
- Sustainable development: Encourage and promote sustainable approaches, ensure the resiliency of the district, and its adaptability to changing natural and human-made conditions.
- Clear identification of the public realm: Develop a generous and interconnected network of open spaces. Also, it can host a variety of activities for the Kabul population.
- Pedestrian and transit-oriented: Design a district at a human scale and promote walkability by capitalizing public transit needs.

2.1 Buildings

The buildings are categorized into several batches based on criteria such as land value, clustering, and budget (Table 1).

Table 1 Building clusters

Batch 1	Batch 2
– Ministry of Finance – Central Bank – Parking	– Supreme Court – Ministry of Economy – Land Independent Authority – National Procurement Authority – Ministry of Urban Development and Housing
Batch 3	**Batch 4**
– Disaster Management Authority – National Bank – Pashtany Bank – Independent Directorate of Local Government – Ministry of Mines and Petroleum – Ministry of Public Work – Masque (Phase-1)	– Capital Region Independent Authority – Ministry of Transport – Parking (P-2 and P-3) – Ministry of Information and Culture – Ministry of Education – Ministry of Refugee and Repatriation
Batch 5	**Batch 6**
– Ministry of Haj and Religious – Ministry of Commerce and Industries – Ministry of Women Affairs – Ministry of Labor and Social Affairs – General Attorney Office – Masque (Phase-2) – Ministry of State for Parliamentary Affairs	– Ministry of Border and Tribal Affairs – General Independent of Kochi – Ministry of Health – Ministry of Counter Narcotics – Parking – Civic Plaza – Fuel liquid Enterprise

3 The Proposed Solution Mechanism

The proposed solution creates a circular transportation corridor that passes through four zones of Kabul and ends at the administrative complex (Fig. 3). An express bus system should be reserved for employees and those who want to go to the complex. This system will reduce air pollution, alleviate traffic congestion from the city center, increase efficient economic, and low energy consumption.

Fig. 3 Access through circular streets to GAC

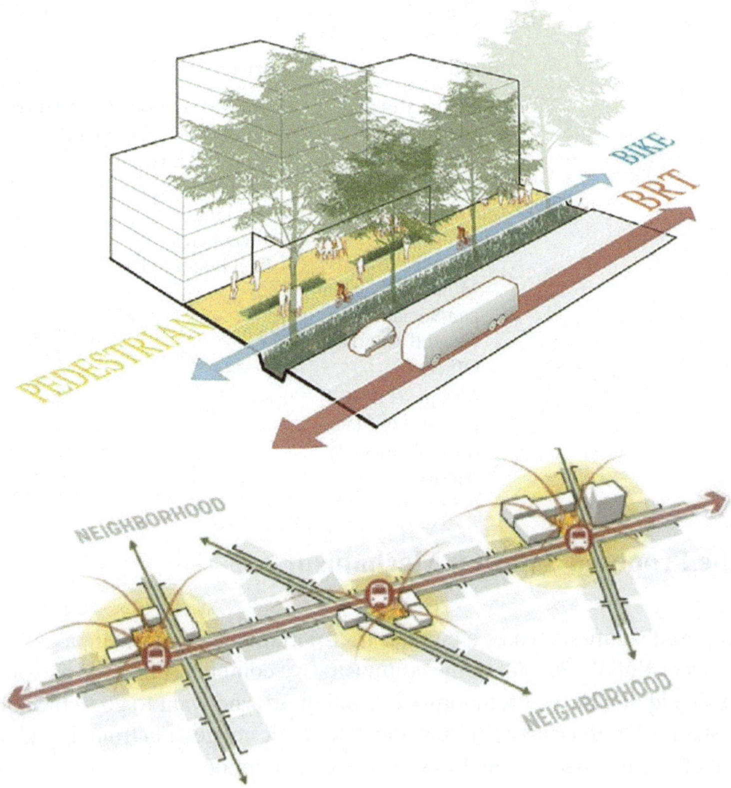

Fig. 4 The proposed BRT, bike, and pedestrian system section [9]

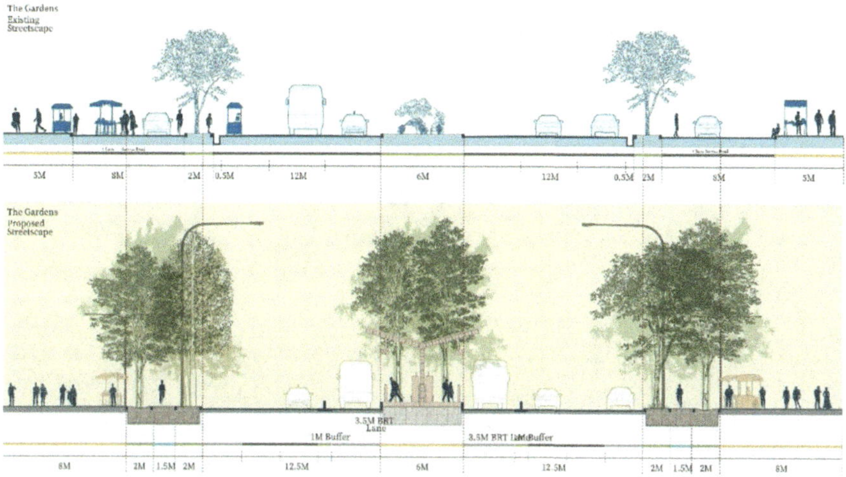

Fig. 5 BRT, bike, and pedestrian system section [9]

Bus rapid transit network: A top priority for Kabul, a city-wide bus rapid transit (BRT) system (Fig. 4), features buses running along dedicated lanes with fixed stations to reduce traffic flow during GAC project construction.

Bike lanes: Along with dedicated BRT lanes (Fig. 5), designated bicycle lanes protected from vehicular traffic are planned. This multimodal network will connect the major city destinations, businesses, and residential neighborhoods, reduce Kabul reliance on private cars, and lead to a more sustainable and equitable city.

Pedestrian walkways: Generous sidewalks will be built close to designated BRT lanes, creating additional medians and safer crossings for pedestrians.

Shared public bicycles: Bicycles are available for shared use on a short term basis for free or a modest rental rate. Bicycles make a statement of the importance of public design.

4 Discussion

4.1 Deployment of the Sustainable Development Goals (SDGs)

By 2030, build the resilience of the poor and those in vulnerable situations, reduce their exposure and vulnerability to climate-related extreme events and other economic, social, and environmental shocks and disasters. According to the Mexico City Climate Action Program (PACCM) 2014–2020, approximately 5.6 million people in the city are vulnerable to the effects of climate change. The Dar Al-Aman GAC addresses SDG goals 3, 4, 6, 9, 11, 13, and 15.

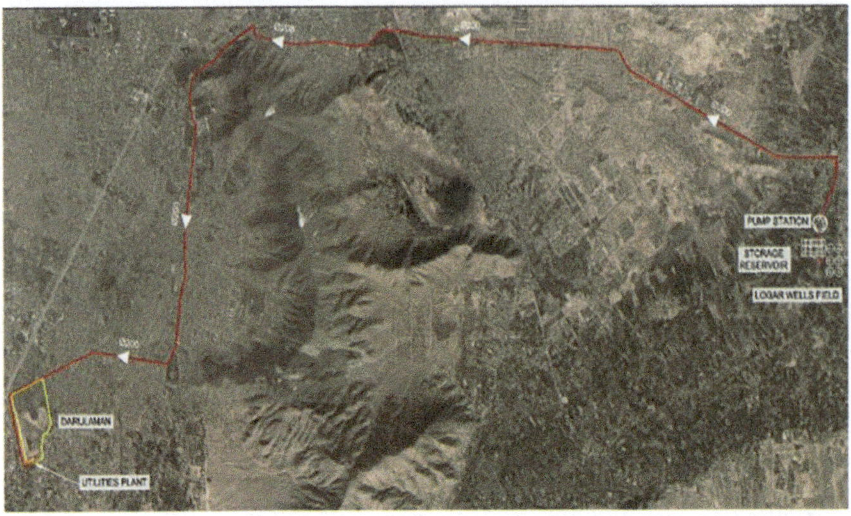

Fig. 6 The proposed water supply from Logar river

Goals 3—Healthy life and well-being (smart life for children): There is concern that the completion of the GAC may cause pollution, population growth, and traffic problems, endangering children's health and well-being. The SDGs state that the government should support a system to improve citizens' health and well-being.

Goal 6—Clean water and sanitation-Smart water management: GAC design includes building pipelines to transfer and pump water from Shatoot Dam and the Logar River well field (Fig. 6). A pipeline with a length of 7.5 km shall be constructed to carry water from the water treatment plant at Shatoot Dam towards a storage reservoir (Fig. 7).

We prefer that gray and black water at Dar Al-Aman GAC be recycled for landscaping use and increase the water table. Besides, absorption of water will be increased by minimizing pavement and other hard surfaces at the site.

Specifications:

- Storage reservoir at Logar wells field having a value of 500 m^3
- Pumping station having the following characteristic
- Flow = 2200 m^3/day—Head = 130 m
- Pipeline of 22 km, having a diameter of 200 mm and a nominal pressure of 16 bars.

Goal 7—Affordable and clean energy: According to reports [11–14], buildings consume 40% of the total energy and emit 30% of greenhouse gases (GHGs) worldwide. Therefore, sustainable energy provision at affordable prices meeting social acceptance, and in an environmentally clean manner [15].

For ensuring sustainability through energy efficiency improvement, 18% for new, and 14% for existing buildings' energy efficiency improvement is suggested [16].

Fig. 7 Proposed water supply system for the Shatoot Dam [10]

Authors in [17] highlighted three options (heating and cooling demand reduction, using energy-efficient equipment, and smartness of the energy systems.

Dar Al-Aman GAC has included diesel generators to supply the ministries' electric power demand, which is not a cost-effective and sustainable solution. SDGs recommends using renewable energy resources to provide a renewable, safe energy source. Using solar panels near the complex can provide affordable and clean energy.

Goal 9—Industry, innovation and infrastructure mobility: Because the government continues to encourage rural citizens to seek opportunities in the city, the population of Kabul has continued to rise. The increased population results in increased traffic, pollution, and poverty.

Goal 11—Sustainable city and communities: Sustainable communities should provide resources to make a city comfortable for residents, but this concept was absent in the Dar-Al Aman GAC.

Goal 12—Smart tourism: As a historical site, Dar Al-Aman is attractive and can attract income as a tourist site. The GAC threatens to restrict access to visitors.

Goals 13—Climate action: The government lacks policies that address climate change through the management of the population, environment, vehicle usage, and urban transportation systems.

4.2 *Leadership in Energy and Environmental Design (LEED)*

Energy efficiency is a key factor for improving design strategy by considering the following actions:

- Increase the quality of insulation.
- Design and purchase higher efficiency-rated appliances and HVAC systems.
- Plan tighter building envelopes.

And consider five principles of an ideal environment:

- Healthy interior environment
- Energy efficiency
- Ecologically friendly materials
- Environment friendly
- Good design
- Sustainable site.

Site selection: An ideal construction site should be chosen to seek LEED goals (Table 2) for better environmental targets for resident place [18]. Site should be situated:

- Away from prime farmland

Table 2 Land use clustering

Land uses	Code	Unit	AM peak hour (7:00–9:00)			PM peak hour (16:00–18:00)		
			Rate	In %	Out %	Rate	% In	% Out
Government	730	Employees	1.02	84	16	–	–	–
Hotel	310	Rooms	0.53	59	41	0.6	51	49
Furnished apartment	220	Dwelling	0.51	20	80	0.62	65	35
Office	710	Employees	0.48	88	12	0.46	17	83
Retail F and B	932	1000 ft^2 gross area	10.81	55	45	9.85	60	40
Museum/exhibition center/library	580	1000 ft^2 gross area	0.28	86	14	0.18	16	84
Common facilities	730	Employees	1.02	84	16	–	–	–
residential	222	Dwelling	0.3	25	75	0.35	61	39

- At 5 feet away from the 100-year flood line
- Away from land inhabited by threatened or endangered species
- At least 100 feet away of any wetlands
- At least 50 feet away from water bodies
- Away from public parkland, unless land of equal or greater value is traded to the public landowner.

The GAC construction site is located on farmland that disturbs agriculture and unsettles the ecosystem.

Site access: Accessibility is one of the most important considerations that determine the success of the GAC project. Visitors from the north and south lose three most important aspects:

- Time
- Money
- Health.

Site design: The RBT system is the most practical proposal that can reduce traffic congestion and pollution. The government should invest in a system of public buses and public bicycles.

- Locate public transportation near trains
- Identify locations for bicycle storage
- Encourage alternative fuel vehicles
- Provide limited parking spaces.

Water efficiency: The water supply for this project is Shatoot dam and Logar, which are high-cost and will take time to complete. Points to consider include:

- Landscaping
- Wastewater treatment
- Water use reduction
- Energy and atmosphere
- Building dimensions
- Renewable energy generation
- Materials and resources
- Renovations for existing buildings
- Waste management
- Reused, recycle or certified materials
- Local or regional materials
- Innovation and design process
- Innovation in design
- LEED accredited professionals.

4.3 *Kabul Urban Design Framework*

The area around and to the south of the Tajbeg Palace contributes to stormwater drainage to the Kabul river. This phenomenon creates an ecological conservation zone with water-sensitive strategies in GAC landscaping. It will support local vegetation and reducing the need for engineered irrigation systems. In the next ten years, more investment will be placed in the southwest part of the city as government agencies consolidate at the GAC. Education linkages should anchor social infrastructure to encourage civic and economic development.

South of Dar-ul-Aman Palace, where hundreds of hectares remain undeveloped and are temporarily used for military and security uses. West of the Palace, a large tract of land has been reserved for future GAC planning and development. This vision has also taken into consideration private parcels with significant transformation potential.

Sasaki sets up the public space framework while establishing several parcels east of Dar-ul-Aman Palace for the construction of the GAC with various accessibility options (Fig. 8). While safeguarding the civic promenade district, these points should be considered:

1. Visionary: A district that represents aspirations of the twenty-first century capital city. Yet it is rooted in the culture and history of Afghanistan and the legacy of the Dar-ul-Aman area.

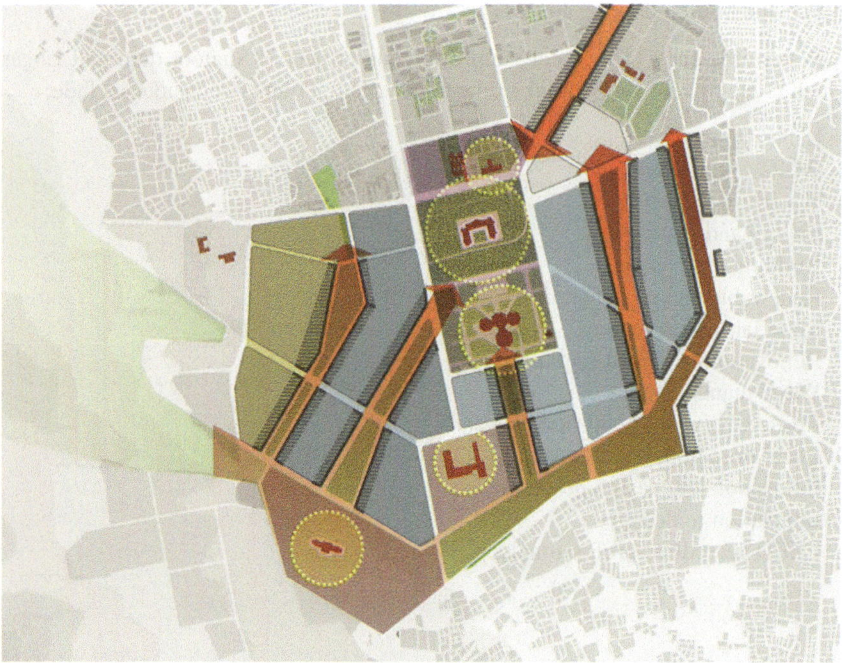

Fig. 8 Accessibility and network to the target place [9]

2. Exemplary: A district that fosters innovative and effective governance and reflects the needs of Afghanistan to connect the government to its citizens.
3. Integrated: A cultural and civic hub with a public space that responds to the natural and cultural contexts, and integrates Dar-ul-Aman Palace, existing ministries, and adjacent neighborhoods and communities.
4. Compact: A transit-oriented development that fosters safe and efficient mobility for all Kabul residents within a compact footprint that preserves land for housing, agriculture, and other uses.
5. Flexible: A forward-looking framework that acknowledges the existing surrounding neighbourhoods and communities with anticipating the evolution of Kabul.
6. Sustainable: An approach that integrates buildings, landscapes, and mobility into systems at every scale to ensure a more resilient Kabul.

Green belt: It seeks to preserve memory and sense of place of the existing site, creating small agricultural areas that can retain jobs for the local community. A neighborhood park provides recreation amenities that serve both residential neighborhoods adjacent to the GAC and government workers (Fig. 9).

Site opportunity: The best opportunity sites locate on the ends of the corridor. At Deh Mazang, a large open area at the junction of the Paghman and Kabul rivers presents a remarkable opportunity for a high impact project. In these areas around Dar-ul-Aman Palace, large publicly owned parcels are currently underutilized. These parcels are privately owned and present great development opportunities.

Integration of larger metropolitan mobility systems is essential to revive the Dar-ul-Aman corridor. Both the JICA Master Plan and the Sasaki Kabul City Framework

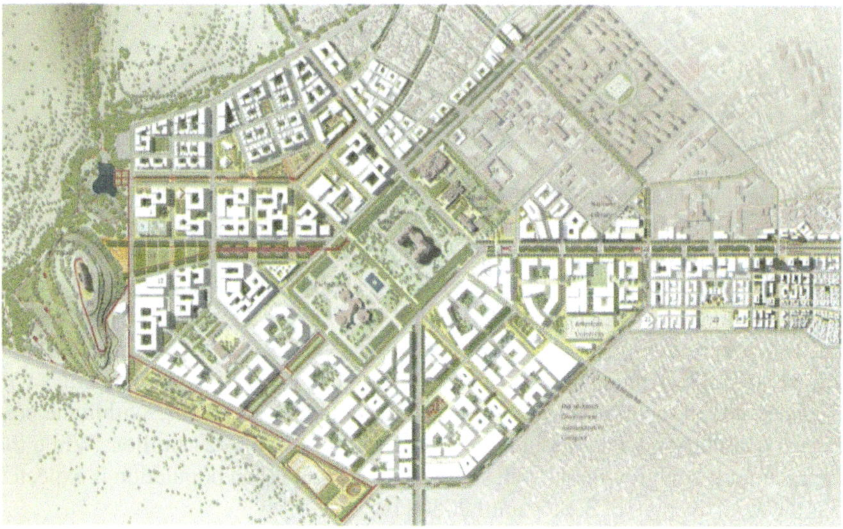

Fig. 9 Green area around GAC [9]

Table 3 Alignment of SDGs, LEED, and urban planning factors in the GAC

	SDGs	LEED	Urban Planning
Aligned	– Decent work and economic growth – Partnership for the goals – Life on land	– Site selection – Building monitoring – Sustainable Site (Site selection, Site Access)	– Street – Landscape – Setbacks – Parking in basement – Stormwater management – Park
Missed	– Good health and well-being – Clean water and sanitation – Affordable and clean energy – Industry innovation and infrastructure – Climate action	– Land within 5 feet the 100-year flood line – Innovation in design – Water use reduction – Renewable energy generation – Building reuse – Reused, recycle or certified materials – Water efficiency (landscaping, wastewater treatment, water use reduction)	– Orientation of building – Solar Panel – Esthetics of forms (form combination) – Public bike – BRT

recommend a city-wide bus rapid transit (BRT) network as one of the top priorities for Kabul in the next decade. For Dar-ul-Aman Road, sustainable and multimodal mobility strategies are crucial given the large-scale relocation of governmental uses from the Civic Hub to the GAC. Critical to this effort is coordination of investments in infrastructure (transit, sanitation, and the like) with public realm improvements (Table 3).

4.4 GAC from Urban Planning Perspective

Afghanistan is undergoing the most powerful wave of urbanization in the past. Cities towns are expanding at rapid rates due to rural–urban migration, influx of IDPs and returnees, and urban sprawl. Afghanistan is still a predominately-rural society with an estimated 24% of the population living in cities [19]. Whereas this trend is changing fast. According to FoAC [19], in 1950, only 1 out of every 20 Afghans lived in cities. In 2014, proportion grew to 1 out of every four, and by 2060, 50% of population will live in cities. Annual growth rate is estimated at around 4%, Afghan cities will expand at one of the highest rates of urbanization in the world [19]. Within the next 35 years, country's urban population is projected to triple to 24 million. While the natural population growth rate will slowly decline, and urban population is expected to grow at an average of 3.14% until 2050 [19].

Cities urban environment consists of natural environment in and around the city, and services/infrastructure environment for the inhabitants. When planning for sustainable cities, a critical consideration of urban environment is potential for climate change, and need for disaster risk reduction [19].

Kabul city has grown vertically and horizontally as land value in the city has risen. Traffic congestion and security issues have motivated the need to create a new government center with easier access away from the city center. This reason can foster coordination and collaboration among different authorities, promote mixed-use with private businesses, and stimulate development of the surrounding neighborhoods. However, new center raises concerns with urban planning near new sites that would destroy arable lands, introduce pollution and congestion, and degrade the historical significance of the site.

Advantages:

- Concentrate on governmental activities.
- Security
- Investment
- Increase land value
- Open space and landscape
- Job opportunities.

Disadvantages:

- Other parts of the city may be neglected
- Access for citizens will be difficult
- Increased population and density
- Underground water pollution
- Threat against preserving cultural heritage.

The case study located in the 7th district of Kabul next to Dar-ul-Aman historical site, the GAC will be the workplace for more than 61,000 employees (Fig. 10). This project will put pressure on managing rush-hour traffic in the 7th and 6th districts. Figure 11 shows access and routes to the GAC.

The light red color shows everyday traffic; bright red color indicates daily traffic in the south part of Kabul. Heavy traffic congestion can waste time and energy consumption, causing pollution problems. Also, orientation and distance between buildings can influence the shadow profiles that could also affect building energy efficiency. The average distance between buildings ranges between 15 and 25 m as average height of the buildings is around 20 m.

4.5 Energy Supply

In lack of renewable supply, power for the GAC will likely be powered by non-renewable resources using one of the following three options:

- One centralized powerplant generator for the entire complex
- Several centralized powerplant generators separately for each cluster

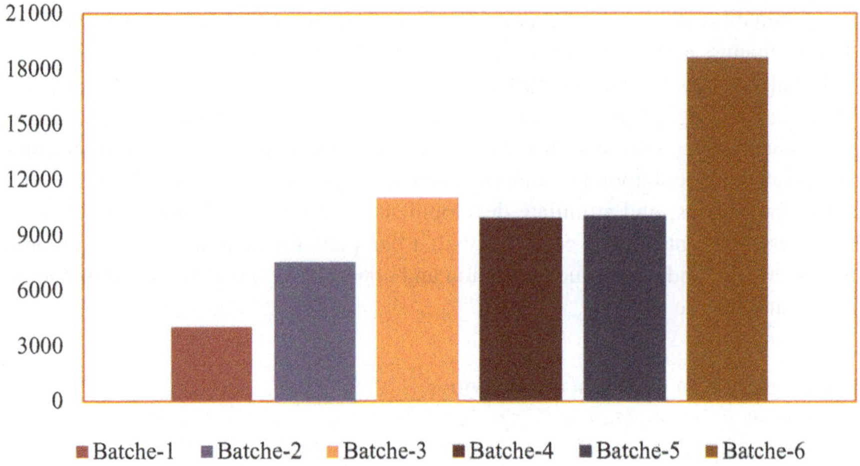

Fig. 10 Number of users that will be benefited by the completion of the GAC project

- Stand-alone backup generator for each building.

Central heating and cooling systems are incorporated into the design, but their operation should ideally be reliable and cost-efficient. As a reliable renewable energy source, solar power can be leveraged to support the GAC because rooftop solar panels may not be feasible to support individual building energy consumption. Construction of a solar thermal plant on the mountainside to provide a backup supply of electricity is suggested. However, rooftop solar panels may be used to address each building demand for water heating. These panels may also facilitate building charging stations throughout the complex.

For traffic congestion solution, three main arterial roads as highly utilized and prone to heavy congestion are suggested. Pollution reduction is feasible by sing electric vehicles and bicycles in the complex.

5 Conclusion

Kabul city is rapidly growing with a population of more than 7 million residents. Heavy-centralized activities at the center of the city aggravate pollution and municipal strain resources. At present, the government administrative offices are scattered and contribute to lack of coordination between these authorities and create accessibility problems for citizens. The government of Afghanistan recently decided to build a central administrative complex and relocate all government offices in one place. This project will be implemented in 3 phases in 100 ha (11.87 million Afghanis) over 10 years. Anticipated advantages to this project include improvement of administrative coordination, concentration of governmental activities and security, increase in

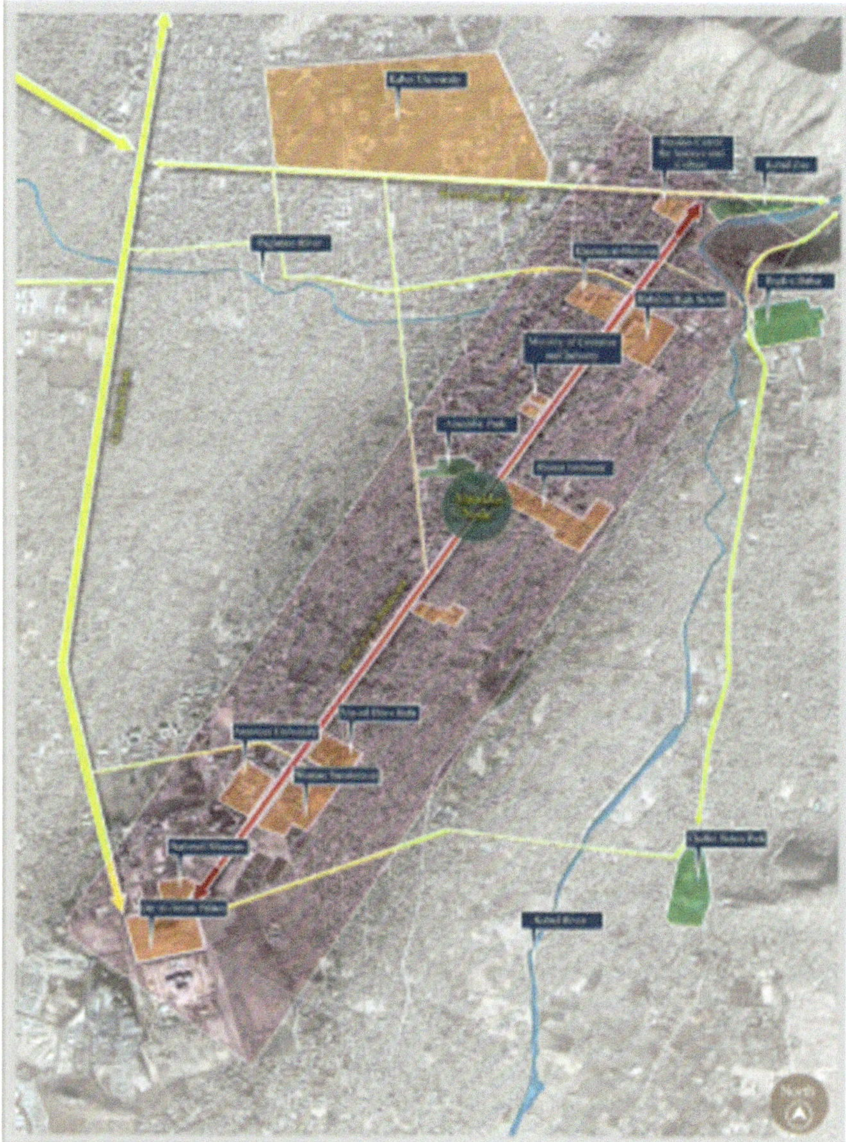

Fig. 11 Access and routes to the GAC [9]

investment, and job opportunity. Disadvantages are accessibility problems, destruction of agriculture land and landscape, reduction of landscape for replenish underground waters, water pollution, neglecting of other parts of the city, destroying of historical identity of the site, and pollution. Case study in this chapter addresses SDGs 8, 15, and 17 goals. Specific lessons-learned in adapt to the Sasaki goals are included as well. The proposed design of the streets, property setbacks, underground parking, and stormwater management, building orientation, solar panels, urban aesthetics, and development of public bicycle and transit are pointed out for further discussion in the future works. This study suggests to regional LEED standards deployment for innovation and design.

References

1. Danish, M.S.S., Senjyu, T., Zaheb, H., Sabory, N.R., Ibrahimi, A.M., Matayoshi, H.: A novel transdisciplinary paradigm for municipal solid waste to energy. J. Clean. Prod. **233**, 880–892 (2019)
2. Danish, M.S.S., Sabory, N.R., Ershad, A.M., Danish, S.M.S., Yona, A., Senjyu, T.: Sustainable architecture and urban planning through exploitation of renewable energy. Int. J. Sustain. Green Energy **6**, 1–7 (2016)
3. Danish, M.S.S., Matayoshi, H., Howlader, H.O.R., Chakraborty, S., Mandal, P., Senjyu, T.: Microgrid planning and design: resilience to sustainability. In: 2019 IEEE PES GTD Grand International Conference and Exposition Asia (GTD Asia), pp. 253–258. IEEE, Bangkok, Thailand (2019). https://doi.org/10.1109/GTDAsia.2019.8716010
4. Danish, M.S.S., Senjyu, T., Ibrahimi, A.M., Ahmadi, M., Howlader, A.M.: A managed framework for energy-efficient building. J. Build. Eng. **21**, 120–128 (2019). https://doi.org/10.1016/j.jobe.2018.10.013
5. Samoylova, N.A., Alekseev, Yu.V.: Urban-panning documentation innovations in sustainable design system. IFAC-PapersOnLine **51**, 780–785 (2018). https://doi.org/10.1016/j.ifacol.2018.11.196
6. Garcia-Garcia, M.J., Christien, L., García-Escalona, E., González-García, C.: Sensitivity of green spaces to the process of urban planning. Three case studies of Madrid (Spain). Cities **100**, 102655 (2020). https://doi.org/10.1016/j.cities.2020.102655
7. Habibzai, A.J.: Afghanistan Green Urban Transport Strategy: 2015–2025 (2015). https://www.afghanengineers.org/wp-content/uploads/2017/01/Afghanistan-Green-Urban-Transport-Strategy-Abdullah-Habibzai.pdf
8. Darulaman Government Administrative Complex Project (2019). https://www.khatibalami.com/360/Angles3.htm
9. Sasaki: Kabul urban design framework. Ministry of Urban Development and Housing, Kabul, Afghanistan (2017)
10. Ministry of Urban Development and Housing (MUDH)—Afghanistan: Shatoot Dam Layout, (2019)
11. Brenna, M., Falvo, M.C., Foiadelli, F., Martirano, L., Poli, D.: Sustainable energy microsystem (SEM): preliminary energy analysis. In: 2012 IEEE PES Innovative Smart Grid Technologies (ISGT), pp. 1–6. IEEE, Washington, DC, USA (2012). https://doi.org/10.1109/ISGT.2012.6175735
12. Bourdeau, M., Zhai, X. qiang, Nefzaoui, E., Guo, X., Chatellier, P.: Modeling and forecasting building energy consumption: a review of data-driven techniques. Sustain. Cities Soc. **48**, 101533 (2019). https://doi.org/10.1016/j.scs.2019.101533
13. Nematchoua, M.K., Yvon, A., Roy, S.E.J., Ralijaona, C.G., Mamiharijaona, R., Razafinjaka, J.N., Tefy, R.: A review on energy consumption in the residential and commercial buildings located in tropical regions of Indian Ocean: a case of Madagascar island. J. Energy Storag. **24**, 100748 (2019). https://doi.org/10.1016/j.est.2019.04.022
14. Kräuchi, P., Dahinden, C., Jurt, D., Wouters, V., Menti, U.P., Steiger, O.: Electricity consumption of building automation. Energy Procedia **122**, 295–300 (2017). https://doi.org/10.1016/j.egypro.2017.07.325

15. Danish, M.S.S., Sabory, N.R., Danish, S.M.S., Senjyu, T., Ludin, G.A., Noorzad, A.S., Yona, A.: Electricity sector transitions in an after war country: a review of Afghanistan's electricity. J. Energy Power Eng. **11** (2017). https://doi.org/10.17265/1934-8975/2017.07.008
16. Danish, M.S.S., Senjyu, T., (eds.): Green building efficiency and sustainability indicators. In: Green Building Management and Smart Automation, pp. 128–145. IGI Global (2020). https://doi.org/10.4018/978-1-5225-9754-4
17. Jensen, P.A., Maslesa, E., Berg, J.B., Thuesen, C.: 10 questions concerning sustainable building renovation. Build. Environ. **143**, 130–137 (2018). https://doi.org/10.1016/j.buildenv.2018.06.051
18. LEED Reference Guide for Building Design and Construction. U.S. Green Building Council, Washington DC., USA (2013)
19. Future of Afghan Cities Programme (FoAC): https://www.fukuoka.unhabitat.org/projects/afghanistan/detail23_en.html. Last accessed 2019/11/01.\

Distributed Generation Model for Achieving Environmental Scenario: Loss Reduction and Efficiency Improvement

Sayed Mir Shah Danish, Atsushi Yona, and Tomonobu Senjyu

1 Introduction

Distributed Generation (DG) has received attention as a solution to environmental and economic challenges caused by conventional power plants. Renewable energy-based DG has rapidly expanded worldwide in recent years because of its promising potential to reduce fossil fuel consumption and reduce power losses and harmful carbon emissions in electrical power generation [1]. Installation of DG impacts system operations and adjusts to the increased load when there is an expansion in the network [2]. Among many investigations have ascertained and implemented these benefits, Stetz et al. [3] discuss the technical and economic benefits of different active and reactive power control strategies for grid-connected photovoltaic (PV) systems in Germany. These strategies control the voltage rise caused by a high local PV power feed-in and hence allow additional capacity to be connected to the main supply. Static phenomena, including voltage drop, network losses, and grid benefits, were analyzed in a high level of grid-connected PV in the middle voltage distribution network by Paatero et al. [4]. Karimi et al. [5] presents a comprehensive consideration of PV penetration on distribution networks. Onlam et al. [6] proposed a novel adaptive optimization algorithm to solve DG placement problems to minimize power loss and improve voltage stability. Hammons [7] examines the integration of renewable energy sources into European power systems, challenges, possible solutions, and development/application of prediction tools for system operation.

In most developing countries, rising demand for electrical energy forces distribution systems to expand and enlarge that may be limited by safety and statutory regulations. When conventional distribution networks are made longer without considering these limitations, there is a risk of an inability to manage unstable voltage safely with

S. M. S. Danish (✉) · A. Yona · T. Senjyu
University of the Ryukyus, Okinawa 9030213, Japan
e-mail: sayedmir.danish@gmail.com

M. S. S. Danish et al. (eds.), *Sustainability Outreach in Developing Countries*,
https://doi.org/10.1007/978-981-15-7179-4_7

consequential large technical and economic losses [8]. Kabul's distribution networks suffer from the above problems. The proposed 20 kV distribution network for Kabul supplies residential, commercial, and industrial consumers with significant problems that include high power demand, low power transfer capability, high power losses, end-user low voltage, high peak load demand, and lack of reserve capacity.

Renewable energy is a promising solution to environmental and economic challenges caused by conventional power plants. Specifically, renewable energy-based DG can reduce fossil energy consumption and harmful carbon emissions and increase electric system reliability and sustainability. The effective integration of these resources into distribution systems relies on their optimal size and placement. In this paper, the new voltage stability index (VSI) [9] can determine the optimal allocation and integration of renewable energy-based DG in radial distribution systems, and the genetic algorithm (GA) determines the optimal size of DG to control voltage deviations and total loss reduction. In this case study, the real-time voltage profile is also simulated for the Kabul distribution network with and without installation of DG's.

2 Problem Statement

The proposed model in this study is located in Kabul, the capital of Afghanistan. Its 20 kV distribution system supplies residential, commercial, and industrial consumers and includes 22 buses and 21 lines, as shown in Fig. 1. (Table 1 lists additional details and parameters.) The targeted network consists of transformer stations (TS) and a Sixth Junction station (JS-6) from the 110/20 kV, 50 MVA Breshna Kot substation. Problems with this distribution network include low end-user voltage and a high level of network losses [10]. As a case study, the efficacy of this distribution network is examined by considering voltage characteristics and losses before and after the integration of DG with renewable energy-based PV.

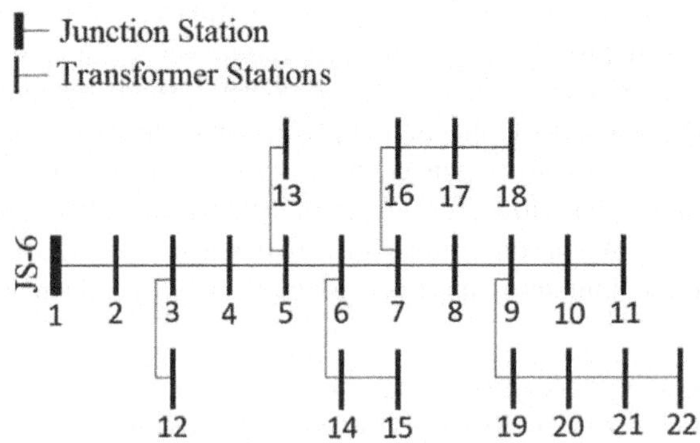

Fig. 1 The Breshna Kot distribution network model

Table 1 The proposed 20-kV distribution system transmission line parameters

Line number	Bus code		Length (km)	R (pu)	X (pu)
	From	To			
1	1	2	0.75	0.2460	0.0724
2	2	3	0.80	0.2624	0.0772
3	3	4	0.60	0.1968	0.0579
4	3	12	0.40	0.1312	0.0386
5	4	5	0.65	0.2132	0.0627
6	5	6	0.95	0.3116	0.0917
7	5	13	0.70	0.2296	0.0676
8	6	7	0.65	0.2132	0.0627
9	6	14	1.40	0.4592	0.1351
10	14	15	0.60	0.1968	0.0579
11	7	8	0.80	0.2624	0.0772
12	7	16	0.65	0.2132	0.0627
13	16	17	0.60	0.1968	0.0579
14	17	18	0.55	0.1804	0.0531
15	8	9	0.65	0.2132	0.0627
16	9	10	0.40	0.1312	0.0386
17	9	19	0.80	0.2624	0.0772
18	19	20	0.45	0.1476	0.0434
19	20	21	0.40	0.1312	0.0386
20	21	22	0.40	0.1312	0.0386
21	10	11	0.45	0.1476	0.0434

3 Voltage Stability Index (VSI) for Optimal DG Placement

The schematic in Fig. 2 shows the concept used to derive the mathematical model from the VSI: specifically, two nodes the source at one end and the load at the other. The mathematical model of the proposed voltage stability index is outlined below. For branch current, I_{12} can be written as follows:

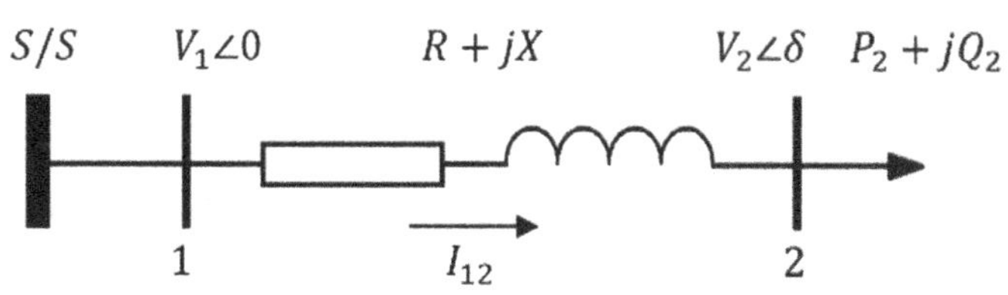

Fig. 2 Simple radial distribution system

$$I_{12} = \left[\frac{P_2 + jQ_2}{V_2\angle\delta}\right]^* \tag{1}$$

The receiving end bus voltage:

$$V_2\angle\delta = V_1\angle 0 - (R + jX)I_{12} \tag{2}$$

Equation (1) can be substituted within Eq. (2) and simplified:

$$\begin{aligned} V_2\angle\delta &= V_1\angle 0 - (R + jX)\left[\frac{P_2 + jQ_2}{V_2\angle\delta}\right]^*, \\ V_2\angle\delta &= V_1\angle 0 - (R + jX)\left[\frac{P_2 - jQ_2}{V_2\angle - \delta}\right] \end{aligned} \tag{3}$$

$$\begin{aligned} V_2^2 &= V_1V_2\angle - \delta - (R + jX)(P_2 - jQ_2) \\ &= V_1V_2\cos\delta - jV_1V_2\sin\delta - (R + jX)(P_2 - jQ_2) \end{aligned} \tag{4}$$

$$V_2^2 + [P_2R + Q_2X + j(P_2X - Q_2R)] = V_1V_2\cos\delta - jV_1V_2\sin\delta \tag{5}$$

The real and imaginary parts in Eq. (5) are calculated for the value $\delta = 0$:

$$V_2^2 + P_2R + Q_2X = V_1V_2\cos\delta,\ for\delta = 0,\ V_2^2 + P_2R + Q_2X = V_1V_2 \tag{6}$$

Equation (7) can be substituted within Eq. (6):

$$V_2^2 + P_2\frac{P_2X}{Q_2} + Q_2X = V_1V_2,\ V_2^2 - V_2V_1 + \left(\frac{P_2^2}{Q_2} + Q_2\right)X = 0 \tag{7}$$

For stable bus voltages, $b^2 - 4ac \geq 0$. VSI is thus defined as Eq. (10):

$$b^2 - 4ac \geq 0,\ V_1^2 - 4\left(\frac{P_2^2}{Q_2} + Q_2\right)X \geq 0,\ 1 \geq \frac{4X}{V_1^2}\left(\frac{P_2^2}{Q_2} + Q_2\right) \tag{8}$$

$$VSI = \frac{4X}{V_1^2}\left(\frac{P_2^2}{Q_2} + Q_2\right) \leq 1 \tag{9}$$

VSI reports three statuses: (I) normal operating conditions when VSI should be less than one, (II) higher stability when VSI near zero, (III) higher instability and vulnerability when VSI has a large value. With this rationale, DG should target the bus with high VSI values. Figure 3 illustrates the voltage stability index profile for the 22-bus network in Fig. 1, showing that the 9th bus would be the optimal placement for a DG node.

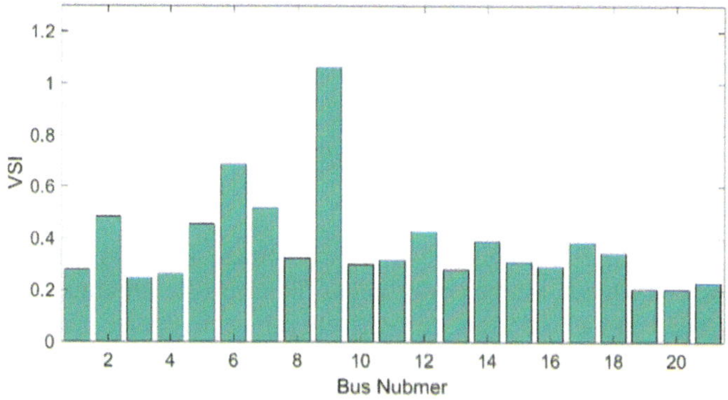

Fig. 3 VSI profile for 22-bus distribution network

4 Optimization and Power Flow Analysis

The Genetic Algorithm (GA) method can determine optimal DG sizes placed in a distribution network [11, 12]. The load flow at this node can be analyzed with the Newton Raphson method to determine the influence of DG's on the network [13]. To analyze power flow, the distribution network operates as a single-phase steady-state model.

4.1 Weighing Factors

DG allocations in the distribution network should minimize power losses and enhance the system voltage profile. The accurate size of the DG to be placed on the vulnerable bus of the network should impact voltage deviation (VD) and active power loss (PL). In order to determine the optimum DG size, the following function is optimized:

$$Minimize F(P_{DG}) = w_1 \times \frac{VD_{DG}}{VD_{BC}} + w_2 \times \frac{PL_{DG}}{PL_{BC}} \tag{10}$$

where w_1 and w_2 are weighting factors of voltage deviation and power losses, respectively; VD_{DG} and PL_{DG} are voltage deviation and power losses as function of P_{DG}, respectively; VD_{BC} and PL_{BC} are voltage deviation and power losses at baseline (before locating DG), respectively; and F is the weighted objective function. Equations (12)–(14) illustrates the mathematical representation of these objectives:

$$w_1 + w_2 = 1 \tag{11}$$

$$VD = \sum_{1}^{N} (V_i - 1)^2 \tag{12}$$

$$PL = \sum_{k=1}^{NB} real\left[Z_k * I_k^2\right] \tag{13}$$

where VD, N, V_i, PL, NB, Z_k and I_k are the voltage deviation index, number of buses, voltage at ith bus, active power losses, number of each brunch, K brunch's impedance and k brunch's current, respectively subject to these constraints:

$$V_{\min} \leq V_i \leq V_{\max} \tag{14}$$

$$P_{sub} + \sum P_{DG} = P_{load} + P_{loss} \tag{15}$$

$$Q_{sub} + \sum Q_{DG} = Q_{load} + Q_{loss} \tag{16}$$

$$P_{DG\min} \leq P_{DGi} \leq P_{DG\max} \tag{17}$$

$$Q_{DG\min} \leq Q_{DGi} \leq Q_{DG\max} \tag{18}$$

The proposed methodology (the Voltage Stability Index method and Genetic Algorithm) are shown in the flowchart shown in Fig. 4.

5 Simulation Results and Discussion

Based on data from all buses and lines modeled upon the physical distribution network infrastructure, VSI determined the optimal location for DG placement at the 9th bus (Fig. 3). Four different cases are presented to show how GA can achieve the optimum size of DG and improve system efficiency.

Baseline Case: The technical parameters of the distribution network consider the real-time daily voltage profile and real network power loss of the network without DG for analytical comparison.

Case 1: The weighting factors are equal (i.e., $w_1 = 0.5$, $w_2 = 0.5$).

Case 2: The power loss weighting factor is lower than the voltage deviation index (i.e. $w_1 = 1$, $w_2 = 0$).

Case 3: The power loss weighting factor is higher than the voltage deviation index (i.e. $w_1 = 0.2$, $w_2 = 0.8$).

The daily residential and industrial load profiles within the distribution network are shown in Fig. 5. Average solar radiation for Kabul during the winter [14], shown

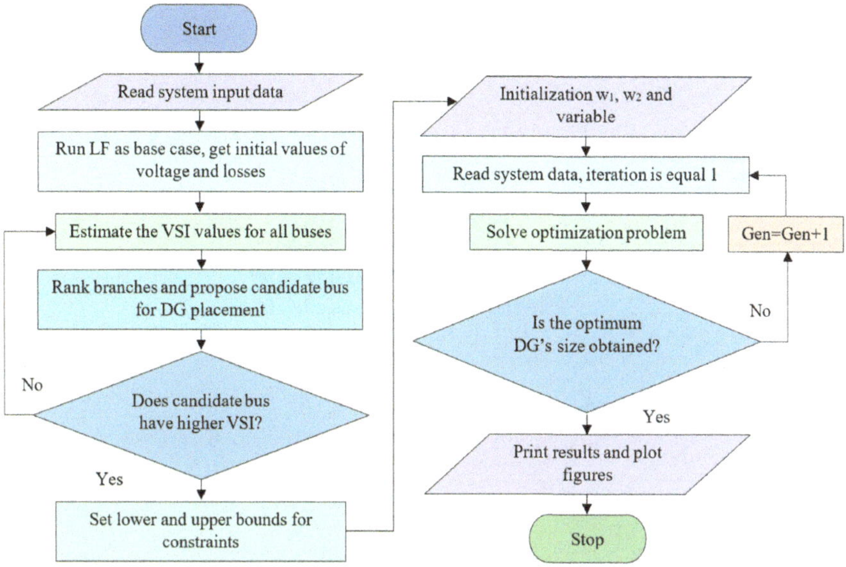

Fig. 4 Flowchart of the genetic algorithm for optimum DG size

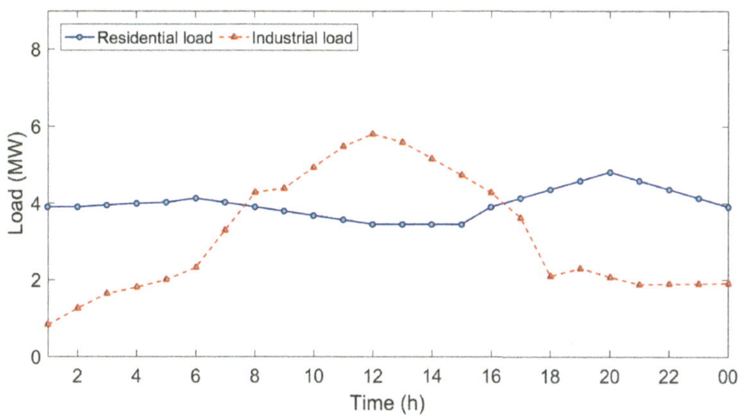

Fig. 5 Daily load profile

in Fig. 6, is used to estimate the daily output power of the PV panels that is integrated with the distribution network.

At baseline, the real-time system voltage profile (without DG) and three cases (with DG) are shown. At baseline (Fig. 7), node voltage deviation exceeds the statutory operational limit (red line) between 7:30 AM and 5:00 PM. For Case 1 (weighing factors are equal), the node voltage profile stays within the statutory limit (Fig. 8). By integrating DG (PDC = 5.0837 MW), the node voltage increases from 0.8571

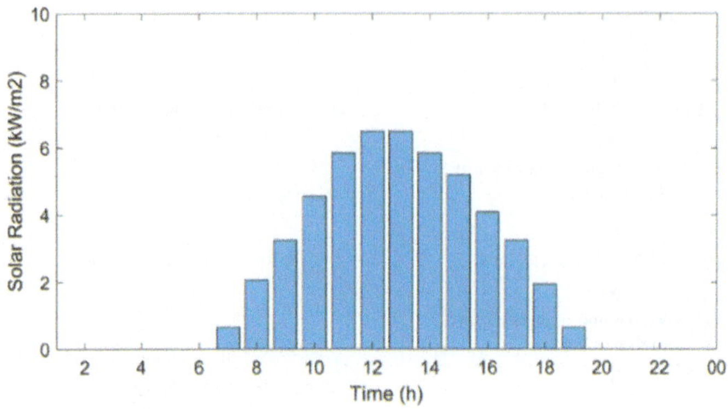

Fig. 6 Average solar radiation

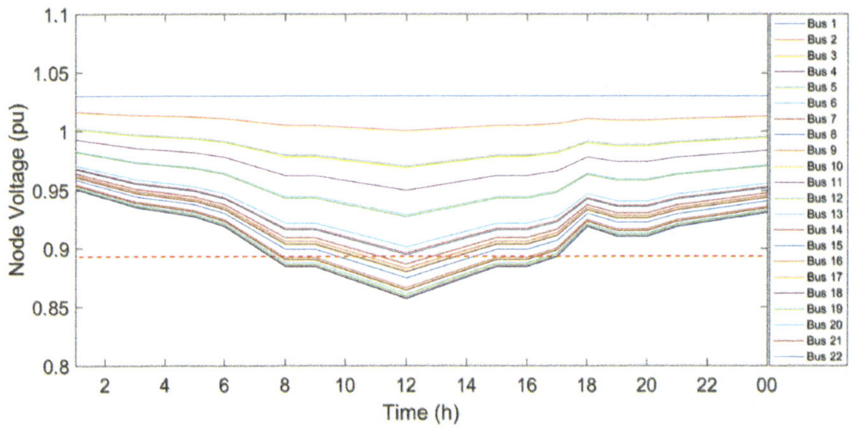

Fig. 7 Voltage profile for baseline case

to 0.9676 pu, which shows 12.892% rise (Table 2). In case 2 (power loss weighting factor is lower than the voltage deviation index), the voltage profile stays within statutory limits with smaller DG size (PDG = 2.0359 MW); the minimum voltage occurs at the peak time (Fig. 9), increasing from 0.8571 to 0.9068 pu. In case 3 (power loss weighting factor is higher than the voltage deviation index), the voltage profile and power loss are greatly improved; the optimum DG size is higher than in other cases. The voltage profile for case 3 (Fig. 10) shows the minimum voltage increased from 0.8571 to 1.0380 pu, a change of 21.106%.

Total active power loss for each case is shown in Fig. 11. Integrating 5.0837 MW (Case 1), 2.0359 MW (Case 2) and 8.1423 MW (Case 3) DG sizes in-network reduces active power loss by 41.512%, 22.402%, and 47.627%, respectively. The best case scenario for the present network is Case 1 (equal weighting factors for voltage deviation and loss reduction); optimum DG size more effectively maintains

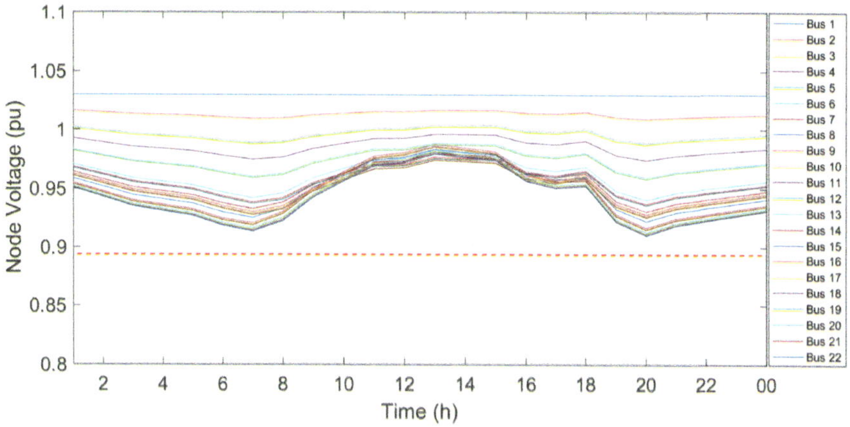

Fig. 8 Voltage profile for Case 1

Table 2 Comprehensive simulation results

	Base case (without DG)	Case 1	Case 2	Case 3
Weighting factor: w_1, w_2		$w_1 = 0.5$, $w_2 = 0.5$	$w_1 = 1$, $w_2 = 0$	$w_1 = 0.2$, $w_2 = 0.8$
DG location (bus number)		9	9	9
DG size (MW)		5.0837	2.0359	8.1423
Daily real power loss (MW)	19.529	11.422	15.152	10.228
Real power loss reduction (%)		41.512	22.409	47.627
Daily reactive power loss (MVAr)	5.7473	3.3616	4.4594	3.0101
Reactive power loss reduction (%)		41.511	22.409	47.625
Peak minimum voltage (pu)	0.8571	0.9676	0.9068	1.0380
Voltage rise (%)		12.892	5.7989	21.106

the voltage deviation within statutory limits and reduces line losses as desired. These results illustrate the benefit of the proposed methodology for DG optimum size and optimal placement in radial distribution networks.

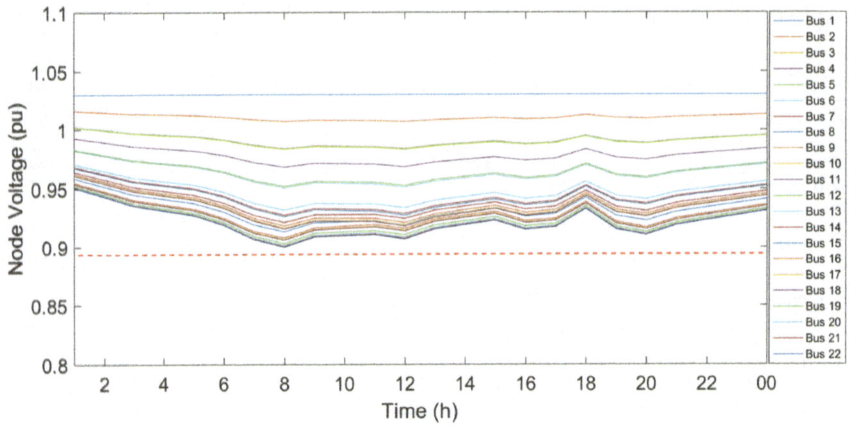

Fig. 9 Voltage profile for Case 2

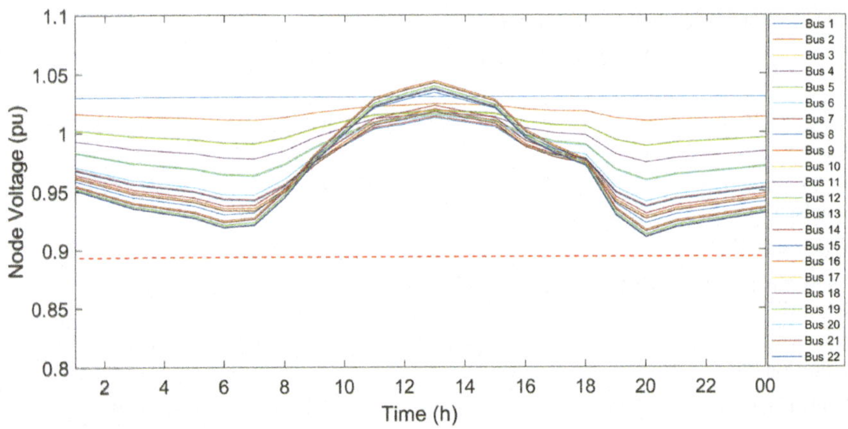

Fig. 10 Voltage profile for Case 3

6 Conclusion

This article is a case study of the 9.7 MVA conventional distribution system located in Kabul. This network experiences low voltage problems, high line losses, and future development constraints under peak load conditions. DG-based renewable energy has received attention as a solution to environmental and economic challenges caused by conventional power plants, and Afghanistan is experiencing significant growth in network-connected renewables, especially at the distribution level. This paper evaluates the effectiveness of DG on voltage deviation and active losses in this network. This study offers tools (VSI and GA) to improve the network reliability by minimizing node voltage deviation and line power losses. Three simulated cases provide

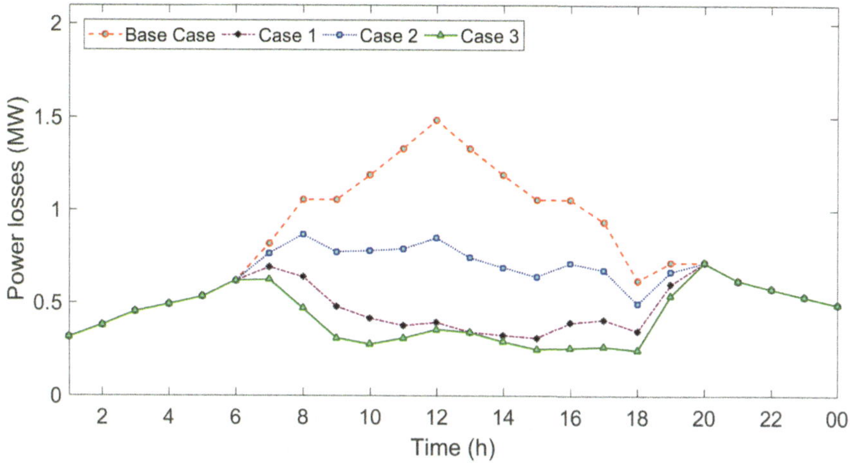

Fig. 11 Total active power losses

an analysis that weighs the importance of voltage regulation and loss reduction as weighting factors. The results show this method can identify the optimum location and size of DG within this network to mitigate problems with voltage regulation and line.

References

1. Ackermann, T., Andersson, G., Söder, L.: Distributed generation: a definition. Electric Power Syst. Res. **57**, 195–204 (2001). https://doi.org/10.1016/S0378-7796(01)00101-8
2. Sultana, U., Khairuddin, A.B., Aman, M.M., Mokhtar, A.S., Zareen, N.: A review of optimum DG placement based on minimization of power losses and voltage stability enhancement of distribution system. Renew. Sustain. Energy Rev. **63**, 363–378 (2016). https://doi.org/10.1016/j.rser.2016.05.056
3. Stetz, T., Marten, F., Braun, M.: Improved Low voltage grid-integration of photovoltaic systems in Germany. IEEE Trans. Sustain. Energy **4**, 534–542 (2013). https://doi.org/10.1109/TSTE.2012.2198925
4. Paatero, J.V., Lund, P.D.: Effects of large-scale photovoltaic power integration on electricity distribution networks. Renew. Energy **32**, 216–234 (2007). https://doi.org/10.1016/j.renene.2006.01.005
5. Karimi, M., Mokhlis, H., Naidu, K., Uddin, S., Bakar, A.H.A.: Photovoltaic penetration issues and impacts in distribution network—a review. Renew. Sustain. Energy Rev. **53**, 594–605 (2016). https://doi.org/10.1016/j.rser.2015.08.042
6. Onlam, A., Yodphet, D., Chatthaworn, R., Surawanitkun, C., Siritaratiwat, A., Khunkitti, P.: Power loss minimization and voltage stability improvement in electrical distribution system via network reconfiguration and distributed generation placement using novel adaptive shuffled frogs leaping algorithm. Energies **12**, 553 (2019). https://doi.org/10.3390/en12030553
7. Hammons, T.J.: Integrating renewable energy sources into European grids. Int. J. Electr. Power Energy Syst. **30**, 462–475 (2008). https://doi.org/10.1016/j.ijepes.2008.04.010

8. Danish, S.M.S., Shigenobu, R., Kinjo, M., Mandal, P., Krishna, N., Hemeida, A.M., Senjyu, T.: A real distribution network voltage regulation incorporating auto-tap-changer pole transformer multiobjective optimization. Appl. Sci. **9**, 2813 (2019). https://doi.org/10.3390/app9142813
9. Murty, V.V.S.N., Kumar, A.: Optimal placement of DG in radial distribution systems based on new voltage stability index under load growth. Int. J. Electr. Power Energy Syst. **69**, 246–256 (2015). https://doi.org/10.1016/j.ijepes.2014.12.080
10. Danish, M.S.S., Senjyu, T., Sabory, N.R., Danish, S.M.S., Ludin, G.A., Noorzad, A.S., Yona, A.: Afghanistan's aspirations for energy independence: water resources and hydropower energy. Renew. Energy **113**, 1276–1287 (2017). https://doi.org/10.1016/j.renene.2017.06.090
11. Wang, Q., Spronck, P., Tracht, R.: An overview of genetic algorithms applied to control engineering problems. In: Proceedings of the 2003 International Conference on Machine Learning and Cybernetics (IEEE Cat. No.03EX693), pp. 1651–1656 Vol.3. IEEE, Xi'an, China (2003). https://doi.org/10.1109/ICMLC.2003.1259761
12. Bansal, R.C.: Optimization methods for electric power systems: an overview. Int. J. Emerg. Electr. Power Syst. **2**, 1–24 (2005). https://doi.org/10.2202/1553-779X.1021
13. Le Nguyen, H.: Newton-Raphson method in complex form [power system load flow analysis]. IEEE Trans. Power Syst. **12**, 1355–1359 (1997). https://doi.org/10.1109/59.630481
14. National Renewable Energy Laboratory (NREL): Afghanistan-NREL Resource Maps and Toolkits. National Renewable Energy Laboratory (NREL) (2011)

Solar Energy Market and Policy Instrument Analysis to Support Sustainable Development

Shawkatullah Shams, Mir Sayed Shah Danish, and Najib Rahman Sabory

1 Introduction

Solar energy is the cleanest and most abundant renewable energy source. It radiates from the sun and can be used for generating electricity, providing light, and heating water [1]. Total installed capacity of solar-based electricity generation has increased from almost negligible capacity in the early nineties to 40 GW in 2010 and 402 GW in 2017 [2]. Solar energy has also experienced an impressive technological shift. In the early years, solar technologies consisted of small-scale photovoltaic (PV) cells. Still, recently these technologies have developed and represented by different types of solar technologies in the world markets [3]. Meanwhile, efficiency and performance of these technologies have increased, and efficiency is approaching 50% in laboratories [4, 5]. Solar energy market has also experienced an extreme expansion due to supportive government policies, increased price of fossil fuels, and concerns with greenhouse gas (GHG) emissions and climate change [3].

Despite the technical potential of solar energy and the recent growth of the market, contribution of solar energy to the global energy supply mix is relatively small. Although, solar energy industry has witnessed a significant decline in capital cost and an increase in fossil fuel prices. At present, solar energy technologies are not yet competitive with conventional technologies for electricity production. Besides the economic disadvantages, solar energy technologies face several technical, institutional, regulatory, social, and cultural barriers towards it large-scale deployment. To overcome these barriers, governments have designed and applied various types

S. Shams (✉) · M. S. S. Danish · N. R. Sabory
Kabul University, Kabul 1006, Afghanistan
e-mail: shawkatullah.shams@gmail.com

M. S. S. Danish · N. R. Sabory
University of the Ryukyus, Okinawa 9030213, Japan

M. S. S. Danish et al. (eds.), *Sustainability Outreach in Developing Countries*,
https://doi.org/10.1007/978-981-15-7179-4_8

of policy instruments such as feed-in-tariff (FiT), renewable portfolio standards, and tax incentives.

The study overviews the current status of the solar energy markets, barriers of solar energy development and deployment, and policy instruments or support mechanisms.

2 Problem Statement and Proposed Solution Mechanism

Solar energy has enormous potential to fulfill global energy demand. Afghanistan is one of the world countries with abundant solar energy resources. According to the Ministry of Energy and Water (MEW) [6], solar energy potential in Afghanistan reaches about 220 GW. Solar energy harnessing will not only help us to become self-sufficient, it will help to become a regional energy supplier [7]. At present, contribution of solar energy supply in Afghanistan's energy grid is still negligible that deployment is slow, facing many barriers. These barriers have economic, technical, and institutional aspects:

- Economic barriers: lack of financing, involvement of high-risk in investment, and high cost of balance of system components (BOS)
- Technical barriers: lack of expertise and experience such as well-trained technicians, engineers, and solar projects implementors
- Institutional barriers: lack of coordination among institutions involved in solar energy or all renewable energy (RE) development, lack of well-defined policies and goals, and lack of financing agencies and research institutes.

This study is performed following these steps: At first, a mixed-method (quantitative and qualitative) analysis was used. A literature review is conducted to determine current status of the renewable energy sector in Afghanistan. This review deals with most relevant publications, reports, and scientific papers dealing with policies encouraging solar and classification of the different types of support (financial, fiscal, political, legislative, and technological). Benchmarking methods for different countries' policies and experiences evaluation is performed. Finally, this study sums up with some measures and policy instruments for successful deployment of solar energy in developing countries. Data and information are collected from related organizations, mainly the Ministry of Energy and Water, and interviews, as well as extracted from research papers, reports, books, and popular websites.

3 Methodology

This research provides an overview of the solar energy market, the main barriers this sector faces, incentive policies deployed, lessons learned, and benchmarks to compare different countries' experiences with solar policies in the world. Objectives of this study are:

- to have a complete and updated picture of different policies and mechanisms encouraging solar energy resources adopted around the world
- to understand reasons behind success or the failure of each policy and mechanism
- to learn lessons for future planning and development of policies and mechanisms
- to identify appropriate supportive measures that help solar energy development in Afghanistan
- to identify the current status of solar energy markets in the world vs. Afghanistan
- to identify main barriers in the way of development and deployment of solar energy technologies, and
- to identify role of policy instruments in the deployment of solar energy technologies in Afghanistan.

This study organized as follows: first, it presents the current status of solar energy technologies, resources, and market expansion over recent years. Then main barriers to development and deployment of solar energy technologies are identified. Finally, it reviews the existing policy instruments used to support solar energy development in developed and developing countries. Lastly, a review of currently existed policy instruments in the energy sector of Afghanistan is performed. Consequently, key findings and conclusions are described.

4 Technologies and Resources

Every day, sun radiates more energy than the world uses in one year [1]. Solar energy technologies can be classified into different categories: passive vs. active, thermal vs. photovoltaic, and concentrating vs. non-concentrating technologies [3]. Passive solar energy technologies collect energy without converting heat or light into other forms. In contrast, active solar energy technologies convert it to other forms for different applications and classified purposes into two parts (photovoltaic (PV) and solar thermal). PV technologies convert radiant energy contained in light quanta into electrical energy (Fig. 1). Mainly, three categories of PV technologies are available in the world markets [8]:

- Wafer-based PV cells
- Thin-film technologies, made from a range of different semiconductor materials, including amorphous silicon, cadmium-telluride, and copper indium gallium di-selenide
- New emerging technologies, hybrid cells, carbon nanotubes cells (CNT), dye-sensitized solar cells (DSSC), and tandem cells/multi-junction solar cells.

Thermal systems use solar heat for thermal, heating, or electricity generation applications. These technologies can be divided into two categories: solar thermal electric and solar thermal non-electric. Solar thermal non-electric category includes applications for agricultural drying, water heaters, air heaters, cooling systems, and cookers. Also known as concentrated solar power (CSP), solar thermal technologies

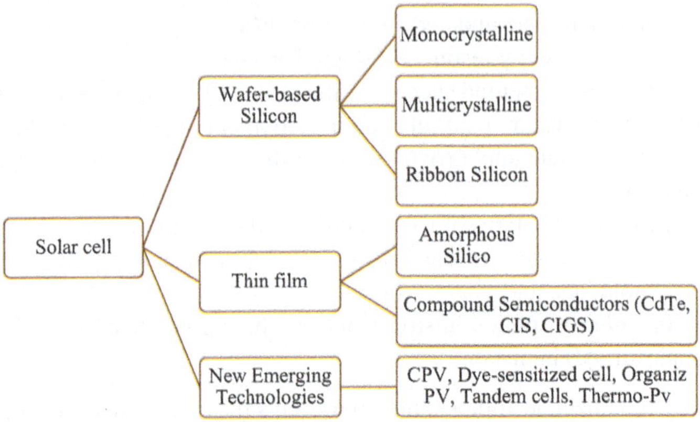

Fig. 1 Classification *of solar cell technologies* [8]

concentrate solar radiation to produce steam or hot air, which can be used for electricity generation using conventional power cycles. Four types of CSP technologies are currently available in the world markets: parabolic trough collectors, Fresnel mirror collectors, solar power towers, and solar dish collectors [3].

Solar energy represents the world's largest source of renewable energy supply, effective solar radiation reaching the earth surface ranges from about 0.06 kW/m^2 at the highest latitudes to 0.25 kW/m^2 at low latitudes. Solar energy in most regions of the world can provide much more energy than the regions consume [3].

Most of Afghanistan lies between a latitude of 30° and 38° north and 60°–72° east with 300 days of sunshine each year. Its average solar potential (Global Horizontal Irradiance—GHI) is estimated about 6.5 kWh per m^2 per day. Higher GHI values are recorded in the southern areas of Kandahar, Helmand, Farah and Herat provinces. Even in the northern provinces, where irradiance averages reaches about 4.5 kWh per m^2 per day, which is feasible for electricity generation. Total solar capacity based on solar radiation and feasible area is reported 222 GW. National average seasonal maximum and minimum GHI are 7.84 and 2.38 kWh/m^2/day [9–11]. Figures 2 and 3 show the GHI and direct normal irradiance for Afghanistan.

Afghanistan should consider CSP technologies for electricity generation. It has 300 sunny days in one year, with annual solar direct irradiance higher than 5 kWh/m^2/day in the southwest part of the country. Figure 4. illustrates monthly average DNI (Direct Normal Irradiance) for CSP technology implementation in major cities. There is an inter-annual variation in normal direct solar radiation. The highest variation can be seen in Herat with a standard deviation of 2.2, and smaller variations in Paktika and Kandahar provinces had 1.06 and 1.23 standard deviation values, respectively [9].

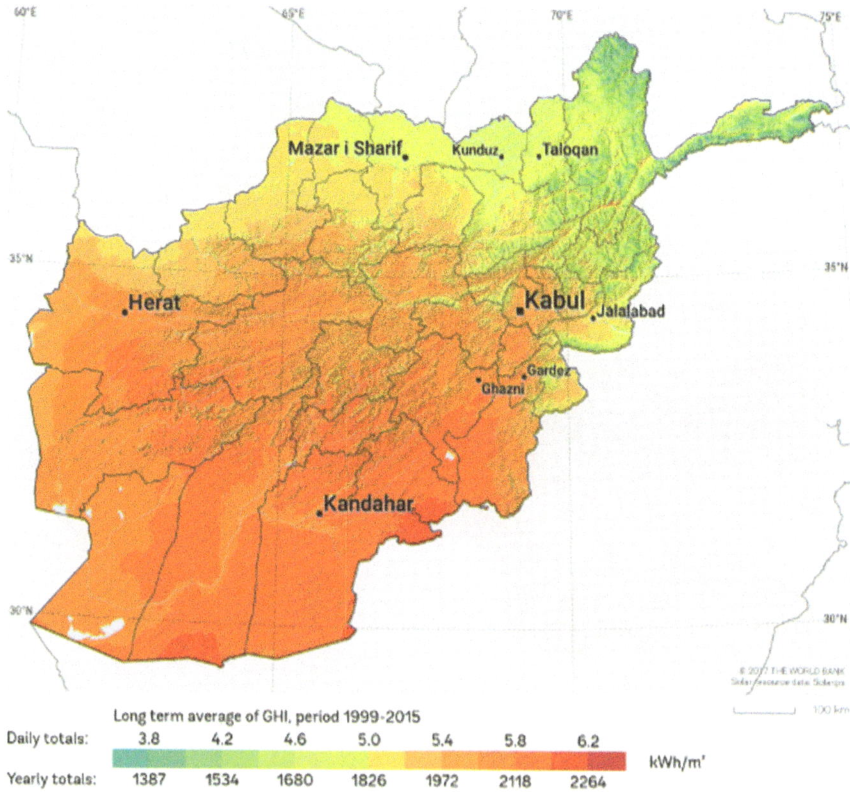

Fig. 2 Global horizontal irradiance for Afghanistan [12]

5 Current Market Status

Installation of solar energy technologies has increased much rapidly in the last years. PV installed capacity has increased from 8 GW in 2007 to 402 GW in 2017 (Fig. 5). CSP installed capacity had increased from 1.095 GW in 2010 to approximately 4.9 GW in 2017. The impetus behind the recent growth of solar technologies is attributed to sustained policy support in countries such as Germany, Italy, United States, Japan, China, India, France, and United Kingdom.

5.1 Solar PV

In 2017, world added more capacity from solar PV compared to other types of power-generating technologies. More global solar PV capacity was installed than the combined net capacity additions of fossil fuels and nuclear power. In 2017, solar

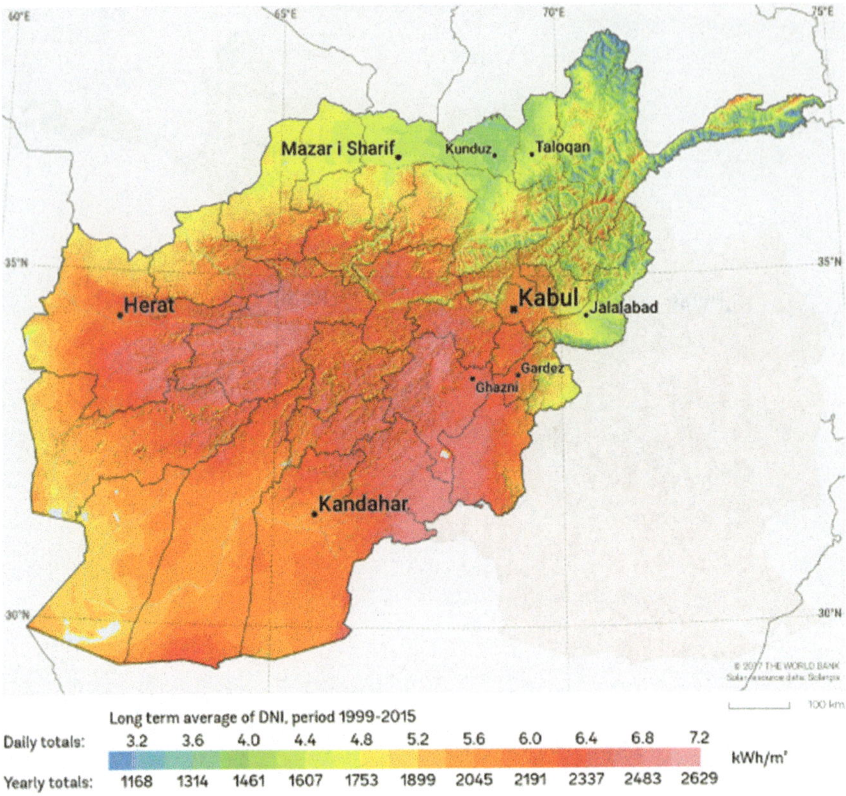

Fig. 3 Direct normal irradiance for Afghanistan [12]

PV was the top source of new power generation source in several major markets, including China, India, Japan, and the United States. On average, the equivalent of more than 40,000 solar panels was installed each hour of the year [2].

By December 2017, global installed capacity of solar PV had reached about 402 GW. Crystalline silicon-based PVs held the major share of the PV market (about 95%). The remainder of the market almost entirely consists of thin-film technologies [2]. Since 2016, China has spurred a significant increase in new solar power installations (up more than 50%). India's market has doubled while other major markets (Japan and the United States) contracted. Top five national markets (China, the United States, India, Japan, and Turkey) were responsible for nearly 84% of newly installed capacity, followed by Germany, Australia, the Republic of Korea, the United Kingdom, and Brazil [2]. Increasing competitiveness of PV, increasing demand for electricity, awareness of solar PV potential for alleviating pollution, and providing energy access are the main reasons behind the expansion of large PV markets globally.

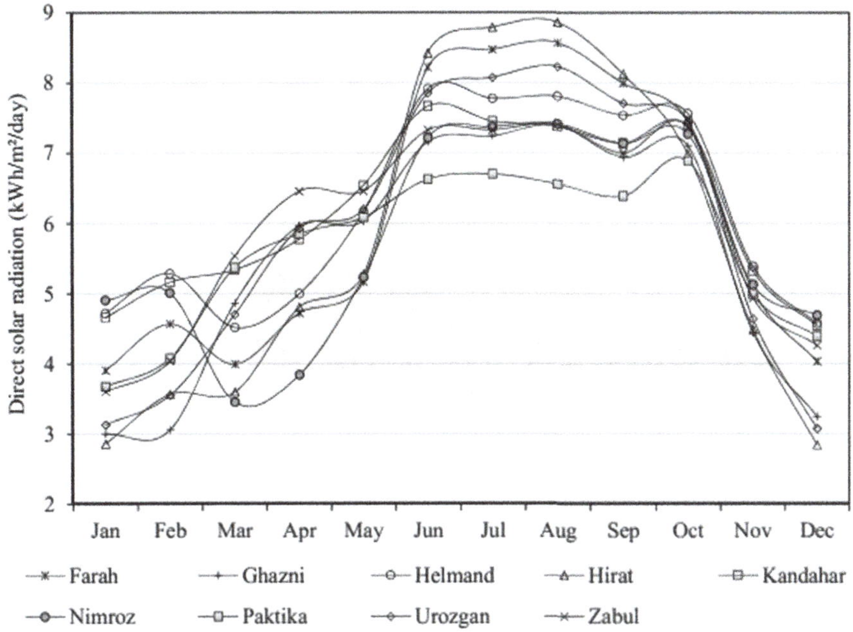

Fig. 4 Monthly average DNI for CSP technologies in major cities [9]

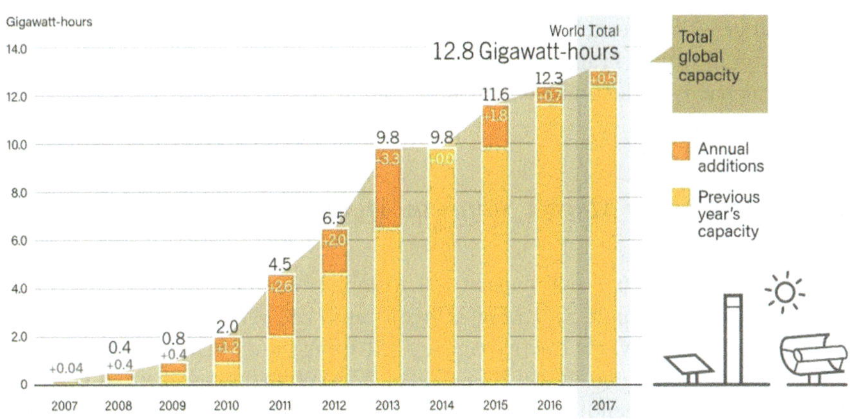

Fig. 5 Solar PV global capacity and annual additions 2007–2017 [2]

5.2 *Concentrated Solar Power (CSP)*

CSP with 100 MW of capacity become available in 2017, bringing global capacity to around 4.9 GW (Fig. 6). Although global capacity of CSP has increased by over 2% in 2017. The CSP industry was active, with about 2 GW of projects under construction,

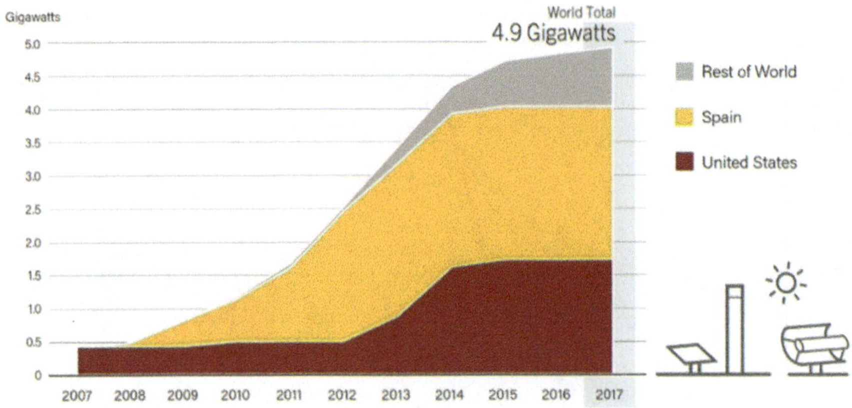

Fig. 6 Concentrated solar thermal power global capacity by country, 2007–2017

particularly in China, the Middle East, and North Africa regions. For the second year running, South Africa led the market in new construction in 2017. Also, for the second consecutive year, new capacity was confined to emerging markets, with no new capacity commissioned in the traditional markets of Spain and the United States [2].

Parabolic and tower technologies continued to dominate markets with approximately 0.9 GW of through systems and 0.8 GW of tower systems under construction by the end of 2017. Fresnel plants totaling about 0.1 GW were at various stages of construction in China, France, and India. India was the only other country in Asia with CSP capacity under construction at the end of 2017 [2].

5.3 Solar Energy Market Status in Afghanistan

According to MEW [6], total capacity of installed renewable energy is 55 MW in Afghanistan, of which 52.9 MW is hydro and 1.8 MW solar. Data provides by International Renewable Energy Agency—IRENA shows that installed capacity of solar energy at the end of 2017 was about 22 MW, including thermal systems. These projects have been financed by international donors, including United States Agency for International Development—USAID, Deutsche Gesellschaft für Internationale Zusammenarbeit—GIZ, Asian Development Bank—ADB, World Bank, Japan Government, New Zealand, and India. According to the energy sector status summary quarterly reports (2016), Afghanistan has 50 MW installed off-grid renewable projects, including 37 MW of micro hydropower plants and 13 MW of solar plants [13, 14].

To meet the targets of the Renewable Energy National Policy [15], which sets a goal to provide 95% energy from renewable energy sources, the government has considered many projects [14, 16]:

- 10 MW on-grid solar PV in Kandahar province
- 2 MW Solar PV + wind hybrid (1.7 MW Solar + 300 kW Wind) project in Herat province
- 35 MW solar PV plant for an industrial park and residential area in the Nangarhar province (currently being analyzed)
- Feasibility study of a 100 MW PV—Hydro hybrid project in Naghlu
- 12 MW solar PV project in Farah province (in the bidding process)
- Feasibility study of a 15 MW rooftop project in Kabul province
- 30 MW solar PV in Kandahar province
- 5.5 MW solar PV and diesel generator hybrid system in Dykondi province
- 400 kW off-grid solar PV plant in Bamiyan province (commissioned in September 2016)
- Solar-wind hybrid project in Herat province
- Tender of a 35 MW solar power plant in the Nangarhar province
- Feasibility study of a 100 MW hybrid project (solar-hydro) next to the existing Naghlu hydropower facility
- Feasibility study of a 10 MW hybrid project (solar and wind)
- Development of a 10 MW solar PV plant with DG backup for Hisar-e-Shahi industrial park in Nangarhar province.

Other projects are under consideration and ongoing at different stages of implementation, feasibility, planning, and surveying. The largest solar thermal system (20 m^3/day) has been installed at women dormitory in Kabul University funded by the World Bank. Other smaller solar thermal systems (domestic and large scale) have been installed in schools, dormitories, children gardens, and military bases since 1980 [6].

6 Barriers to the Development and Deployment of Solar Energy

Since the beginning of the 21st Century, renewable energy has been a significant area of research among scientists. Although scientists and scholars have developed innovative technologies that capture renewable energy, and convincing public to switch from using non-renewable energy sources with quite slow trends, especially in the developing nations. To ensure a sustainable energy future, energy sector must be accorded with priorities. Efforts to reduce reliance on fossil fuels by increasing share of renewable energy in the energy grids have met with little success.

6.1 Technical Barriers

Although PV technology is mature and has improved enormously in the last decade, there are still numerous technical barriers that PV technologies have faced:

- Efficiency constraint: 4–12% (for thin-film) and under 22% (for crystalline) in the available markets [3].
- Performance limitations of the balance of system (BOS) components such as batteries, inverters, and other power conditioning equipment [17].
- Availability of cadmium and tellurium for certain thin-film cells, which are by-products from zinc and copper processing [3].

For solar thermal technologies, following technical barriers are highlighted:

- Limitations to overall system efficiency, such as heat-carrying capacity of transfer fluids, thermal losses, and storage issues with CSP. For central receiver systems, new energy storage technologies such as molten salt-in-tube receiver technology and volumetric air receiver technology, remain unproven for large-scale application [3].
- Limited orientation and design of solar water heaters have not yet matched diverse consumer demand profiles.

6.2 Economic Barriers

Initial capital cost, high risk of investment, and high cost of balance of system (BOS) are the main barriers to solar PV technologies. High initial capital cost with the lack of financing options is the primary barrier in developing countries. Second, lack of financial institutions' experience in investment in these projects while considering a high-risk investment. Finally, the BOS cost decline is not parallel with the modules' prices. Costs of different solar modules are decreasing day-by-day, but cost of other necessary items associated with PV system remains unchanged. So, it leads to a hike in the overall cost to establish a solar PV plant. Furthermore, overall costs with fossil fuel technologies are lower than renewable energy technologies due to currently subsidized operational costs [18–20].

Solar thermal systems face economic barriers with credit risk, costs for requisite backup systems, and high cost of solar thermal water heating system. First of all, a high initial down payment with long payback terms and anticipated mall revenue streams creates a high creditworthiness risk. Backup heater systems essential for a water heating system increase the overall cost. Although cost of domestic water heating systems is much lower than the cost of solar thermal-based water heating systems. Consequently, creditworthiness, backup systems, and high costs made these systems unaffordable [3].

6.3 Institutional Barriers

Institutional barriers include inadequate understanding, procedural problems, and grid access restrictions. National and local institutions should better inform public about the advantages of sustainable energy. Education and training programs for workforce should be provided to build, maintain, improve, and monitor performances of solar facilities. Procedural problems need to be streamlined among public sector agencies [21, 22]. Also, problem with renewable technologies integration in national grid for interconnection, metering, and billing should be solved. More effective policies, such as renewable portfolio standards (RPS) for utilities are needed to encourage wider adoption and understanding among key national and local stakeholders [3].

Private sector investment in solar energy or other renewable energy projects in some countries is hindered by the lack of well-designed policies and approval process delays [18]. Because policy-makers have not been able to develop incentives for private investment in renewable energy or streamlined approval processes with sustainable energy projects, large-scale solar energy or renewable energy projects remain unattractive.

7 Policy Instruments that Support Solar Energy Development in the World and Afghanistan

Since solar energy technologies are not yet cost-competitive with conventional energy commodities at either wholesale or retail levels, significant deployment of solar energy will not be possible unless major policy incentives are introduced. Consequently, many governments have passed a broad range of fiscal, regulatory, market, and other instruments to support solar energy development. Several studies are analyzed various policy instruments, including solar at global and national levels. Viable policies have sustained growth in solar energy markets in Europe, United States, and some developing countries like India.

7.1 Financial Aspects

Since solar energy projects have higher capital costs, therefore these projects require financial incentives for implementation. Technology and resource uncertainties invest in these projects riskier [23]. As developing renewable energy technologies (RETs) are proven, the risks of investing in RE projects will be reduced. In countries where research and development for RETs have been well established, more financial institutions are willing to provide favorable loans to invest in RE projects. Governments also play an early role in early stages of technology development by creating necessary framework and conditions to encourage initial contribution and investment [23].

Therefore, financial support, political commitment, and technological innovation are essential for successful integration of RE technologies. Grants and loans from public or private sources can cover capital or operational costs.

- Public sector funding:

One of the main drivers of solar energy development in developing countries is public sector/government investment. Many developing countries support several government and donors funded projects for solar energy for rural electrification programs [23]. This support can provide grants, low-interest loans, and loan guarantees [5]. These incentives reduce burden of the initial investment by decreasing equipment costs and address market barriers by reducing investment risk.

- Private sector funding:

Private sector funds of RE projects come in form of structural funds or favorable loans from banks and other financial institutions such as venture capital organizations [23].

- Structural funds:

European Cohesion Policy created the financial instrument known as "structural funds" to achieve the European Union (EU)'s renewable energy and energy efficiency targets. It provides support for project development, training, and other key measures designed to reduce unemployment and stimulate economic activity.

- Favorable loans:

Access to funding, especially for smaller companies, is needed to fund pre-feasibility studies for RE projects. Banks or private sector financial institutions can issue preferential loans and guarantees or favorable (low interest) loans to support these analyses [24].

7.2 *Fiscal Aspects*

The cost of energy from renewable energy resources is higher compared to conventional sources such as fossil fuels and nuclear power plants. Capital costs for building conventional power plants are usually subsidized, which is not the case for solar energy or RE power plants. However, costs with using fossil or nuclear fuels do not include additional costs associated with eliminating wastes or pollutants from the air and water. Disposal of nuclear wastes also brings risks to radioactive contamination. Some fiscal support is needed to support RE [23]:

- Positive budgetary incentives to encourage investment in solar energy or RE.
- Environmental tax levies, which penalize the use of fossil fuel.

7.3 Legislative Aspects

Since solar energy market is a policy-driven market, clear legal frameworks are needed to increase investors interest. Because delivering RE and solar energy cost more than other forms of energy. Still, concerns about the financial viability of many RE projects have remained. Well-designed legislative instruments can help develop RE markets [23] by legislation overseeing the purchase of energy, and regulations and mechanisms that facilitate energy grid access.

- Power purchase legislation:

Different types of legislative mechanisms for power purchase regulations are available. These mechanisms (Feed-in tariffs (FiT), tendering arrangements, renewable portfolio standards (RPS), and green certificates, and green pricing schemes) can be adapted to provide support for guaranteed solar energy and RE markets [23].

- Grid access legislation:

Integration of solar energy and RE technologies to grids is essential for successful implementation. Because RE projects are generally small scale, decentralized, and may be located in rural or remote locations, grid connections are limited or unavailable. Achieving the highest level of renewable energy penetration requires successful grid integration [25].

7.4 Political Aspects

Local and national political support are required a successful diffusion and penetration of solar energy or all RE. A well-defined, strong and long-established policy should establish a regulatory framework and price support mechanisms to support solar energy projects and ensure funds for national research and development programs. Adopting official targets for solar energy levels in domestic energy mix is a prominent example [23]. By establishing a target, governments send an important political message that encourages deployment of solar energy. Reaching the goal increases market share of renewables resulting in decreased costs with increased production and a larger infrastructure base. National energy policies aim to encourage diversity and security of supply, to reduce imports of fuels, and to reduce greenhouse gas emissions. RE can make an important contribution towards achieving these objectives [3, 25].

7.5 Technological and Environmental Aspects

Development of renewable energy technologies requires support at all stages of research and implementation to build competitive local industry capabilities in RE. Some renewable energy technologies at an early stage of development, such as CSP, need support to overcome technological and economic barriers. These programs reduce capital and operating costs, improve efficiency, and demonstrate long-term reliability of the technology. Technological supports as new and mature technologies need to address storage, intermittence, cooling, and cleaning challenges. Greenhouse gas emissions and climate change have also received international attention. Countries are asked to reduce their "carbon footprint" generate from coal, gas thermal, and gas combined cycle. This environmental awareness drives technological support and development of RE projects [23].

8 Use of Policy Instruments in Developed and Developing Countries

Developed and developing countries have designed and implemented different types of price-based and quota-based mechanisms to promote RE development. Developed countries introduced these processes from late 1970s. United States implemented its first FITP in 1978 [26], and a quota mechanism is known as renewable portfolio standard (RPS) from 1983; today 31 of 50 states have these instruments in place. Germany was the first European country to introduce a feed-in tariff (FIT) in 1990. Since then, many European countries have implemented either price-based or quota-based mechanisms [25]. Developing countries also have a long history of designing and implementing regulatory instruments to promote RE. The first four countries that introduced some type of preferential tariff or FIT were India (from 1993), Sri Lanka (from 1997), and Brazil and Indonesia (from 2002) [25].

Quantity-based mechanisms, however, have been less popular in the developing world. If a specific target is specified in a rigorous RPS, then utility companies will begin to set competitive goals to meet those targets, as shown in countries like Chile (from 2008), Poland (from 2005), and Romania (from 2004). Although the use of competitive schemes or auctions to deploy RE in the developing world is less common, some countries have or are now testing their effectiveness (Argentina, Brazil, China, Peru, Thailand, and Uruguay). Today, FITPs are being implemented in more than 49 countries around the world and are often cited as the most effective policy for attracting private investment in RE. However, many developed and developing countries still use quantity-based mechanisms, including RPS and auctions (for example, Brazil, Chile, China, France, Poland, Sweden, the United Kingdom, and the United States). Other supplementary measures directly stimulate investment in RE. These have been adopted in parallel to price- and quantity-setting instruments

[25]. A few developed countries have also been applying carbon taxes since the beginning of the 1990s (Netherlands and Scandinavian countries). While other countries have only recently started to apply them (Canadian Province of British Columbia). As of today, no developing country has formally implemented a greenhouse gas cap-and-trade scheme or a carbon tax [25].

9 Policy Challenges

Penetration and deployment of solar energy are largely facilitated by government policies and incentives. The rapid market growth of solar energy in Germany and Spain could be attributed to the feed-in-tariff systems that guarantee attractive returns on investment and to regulatory requirements mandating 100% grid access and power purchase. Government incentives and regulatory mechanisms such as RPS facilitated rapid deployment of solar energy in the United States. Having an appropriate policy instrument is necessary to surmount obstacles and challenges. Some issues like decreasing government revenues, or the slow pace of technological development can put these instruments in a transitional phase. For instance, in Germany, FIT rate is being reduced and could drive investors away from the solar energy market [3].

Cost sensitivity affects decisions and commitments to solar power in developing countries. India planned in its eleventh Five-Year plan (2007–2012) to install 15,000 MW of grid-connected renewable energy driven by wind, micro-hydro, and biomass. This plan recognizes that solar PV would be an option only if the prices come down to levels comparable to micro-hydro [3]. After launching the National Solar Mission to promote solar power in India, the government set first-phase (2009–2013) targets to increase the utility grid power from solar sources, including CSP, by over a 1 GW [25]. By 2022, 20 GW of solar capacity should be added in India. Similar approaches to add solar power to renewable energy commitments are also encouraged in China, Philippines, and Bangladesh. In Brazil, technology-specific and reserve energy auctions ensure that the cheapest renewable energy projects are implemented first [3].

Because solar PV can provide electricity to remote, rural areas that do not have access to the grid, some countries have made developed rural electrification programs, for example, solar home systems (SHS). These solutions require significant subsidies since rural areas are characterized by low-income households that may not be able to afford solar energy technologies [3].

Providing subsidies in the form of government funds or through international donors should be considered short-term support, not a long-term solution. While CSP and solar water heating could be cost-competitive with conventional fuels. Fossil fuel subsidies are politically sensitive in many countries, and their removal might take time. To date, CSP has not been very successful in developing countries. CSP is limited to utility-scale applications and is much more expensive compared to the traditional utility generation markets. Developing countries have adopted a cautious policy approach to solar power, focusing more on pilot-scale projects and grid-connected solar. India is the first developing country to take a step toward installation of CSP [3]. Unlike other electric applications, solar heating applications enjoy limited policy support since instruments like FITs and RPS do not apply to heating applications. Moreover, it is more difficult to measure and verify solar water heating performance, so performance-based incentives are harder to enact.

10 Policy Mix

A mix/suite of policies can grow solar energy markets by guaranteeing returns on investment. In the United States, a policy portfolio that includes federal tax credits, subsidies, RPS, net metering, and renewable energy certificates facilitated solar energy market growth. A 354 MW of parabolic trough CSP in California was supported by a combination of federal tax credits, favorable utility power purchase agreements, and property tax exemptions from the state [3].

11 Discussion and Recommendations

Once energy self-sufficient, Afghanistan has the potential to export energy to neighboring countries. To harness and deploy solar energy, a well-designed, clear and simple policy instrument package is necessary. Afghanistan government has drafted some policies but not yet legislatively approved, such as FiT, net metering, and power purchase agreement for solar energy development and deployment. As a result, no well-defined policy instruments exist that can address all of the financial, fiscal, technological, and legislative concerns.

11.1 Financial Aspects

Financing mechanisms play a key role in encouraging investment in solar energy projects by decreasing equipment costs and addressing market barriers. The following measures should have a positive impact on development of the solar energy markets:

- Public sector funding, such as loans and grants from the government, to reduce the burden of the initial investment to overcome economic market barriers.
- Research and training infrastructures funding to address technical and investment risk barriers.
- Private sector engagement and investment in solar energy projects.
- Capital subsidies.

11.2 Fiscal Aspects

Currently, there are no clear budgetary support mechanisms for sustaining solar energy development in Afghanistan. The following measure and policy instruments are suggested to develop positive fiscal arrangements for solar power consumers and investors:

- Income tax exemption or at least reduction, for example, income tax exemption or reduction for 5 or 10 years.
- Import duty reduction or exemption on solar energy technologies, especially PV system components.
- Tax credits for investment or production to reduce costs and encourage investment in the solar energy market.

11.3 Political Aspects

A consistent commitment from the government is crucial for successful penetration of RE. An overall national plan for RE implementation includes not only a regulatory framework and price support mechanisms but also domestic investment and grants in research and development programs. An important political message that encourages the deployment of RE is the adoption of milestones and targets for implementation. By reaching these goals, the market share of renewables in domestic energy use will increase, and costs will drop with increased production and a larger infrastructure base.

11.4 Legislative Aspects

The Ministry of Energy and Water (MEW) has drafted legislation for support mechanisms for solar energy development such as FiT, net metering, and power purchase agreements.

- The support mechanisms should be approved quickly as the lack of any clear policy causes uncertainties and increases investment risk.
- Some support mechanisms, like the renewable portfolio standard (RPS) should be considered for the development and deployment of solar energy.

These instruments create continuous incentives for energy providers by establishing cost competition among renewable producers for their share of the RPS and for securing contracts. These transactions do not involve the government budget. These additional points should be considered during the design and application of policy instruments:

- Choice of policy instruments, policy design, and complexity of the policy package or regulatory regime should be designed with respect to the actual conditions of the system in the type of market.
- Effectiveness of RE policy depends on the sequencing of implementation to ensure regulatory effectiveness and administrative efficiency.
- Consider parallel systems while policies are designed and evaluated, such as a clear rule for transmission and grid access during implementation.
- A close coordination among institutions involved in RE development is required. For feedback on improving design and implementation of policies and support mechanisms.
- A strategic portfolio/mix/package approach may be more useful to achieve the overall objective for RE development rather than considering policies individually.
- In countries where RETs and research and development laboratories are well established, more financial institutions are willing to provide favorable loans for the RE project implementation.

12 Conclusion

As the cleanest and most abundant renewable energy source available, solar power can supply the entire global energy demand for a variety of uses, including generating electricity, providing light, and heating water for domestic, commercial, or industrial use. Afghanistan has great solar energy potential, with 300 sunny days and an average GHI of about 6.5 kWh per m^2 per day in southern areas and about 4.5 kWh per m^2 per day in northern areas to generate about 222 GW. Afghanistan can also provide concentrated solar power (CSP), with direct solar irradiance higher than 5 KWh per m^2 per day in the southwest part of the country. Despite this technical potential and the recent growth of the market, contribution of solar energy to the total energy supply mix is still small. However, energy organizations in Afghanistan have encouraged implementing support mechanisms such as FiT, PPA, and net metering to support solar energy development, currently being drafted.

While solar power generation demand has increased worldwide, countries strive to reach goals for emission reduction and renewable power generation. Therefore, to meet future energy demand and to provide sustainable energy solutions, policy instruments are known essential. However, solar energy has faced technical, economic, institutional, policy, regulatory, and social/cultural barriers that have impeded technology diffusion in developed and developing countries. Since solar energy technologies are not cost-competitive with conventional energy technologies, large-scale deployment of solar energy is not possible without introducing significant incentives. A large number of support mechanisms or policy instruments are supposed (such as FiT, Renewable portfolio standards, tax incentives, etc.) to overcome barriers and develop solar energy at the national level.

References

1. Breeze, P.: An introduction to solar power. In: Solar Power Generation, pp. 1–4. Academic Press, Massachusetts, United States (2016)
2. Renewable energy 2018: Global status report. Renewables Now (REN21), Paris, France (2018)
3. Timilsina, G.R., Kurdgelashvili, L., Narbel, P.A.: Solar energy: markets, economics and policies. Renew. Sustain. Energy Rev. **16**, 449–465 (2012). https://doi.org/10.1016/j.rser.2011.08.009
4. Annual Report 2016/17. Fraunhofer Institute for Solar Energy Systems (ISE), Berlin, Germany (2017)
5. Unlocking renewable energy investment: The role of risk mitigation and structured finance. International Renewable Energy Agency (IRENA), Abu Dhabi, UAE (2016)
6. Ministry of Energy and Water (MEW)—Afghanistan: Afghanistan Agrometeorological Bulleting (2018). https://mew.gov.af/
7. Sharafmal, F.: Solar Energy Status in Afghanistan (2018)
8. Mohanty, P., Muneer, T., Gago, E.J., Kotak, Y.: Solar radiation fundamentals and PV system components. In: Mohanty, P., Muneer, T., and Kolhe, M., (eds.) Solar Photovoltaic System Applications: A Guidebook for Off-Grid Electrification, pp. 7–47. Springer International Publishing, Cham (2016). https://doi.org/10.1007/978-3-319-14663-8_2
9. Anwarzai, M.A.: Research and Analysis of Afghanistan's Wind, Solar, and Geothermal Resources Potential (2018). https://tuat.repo.nii.ac.jp/?action=repository_action_common_download&item_id=1487&item_no=1&attribute_id=16&file_no=1
10. Ershad, A.M.: Institutional and policy assessment of renewable energy sector in Afghanistan. J. Renew. Energy **2017**, 5723152 (2017). https://doi.org/10.1155/2017/5723152
11. Ershad, A.M., Brecha, R.J., Hallinan, K.: Analysis of solar photovoltaic and wind power potential in Afghanistan. Renew. Energy **85**, 445–453 (2016). https://doi.org/10.1016/j.renene.2015.06.067
12. Solar resource maps of Afghanistan. https://solargis.com/maps-and-gis-data/download/afghanistan. Last accessed 2019/11/01
13. Resource Assessment: Renewable Energy Capacity. https://public.tableau.com/views/IRENAREsourceRenewableEnergyCapacity_CountryProfile/Charts?:embed=y&:showVizHome=no&:size=35,1&:embed=y&:showVizHome=n&:bootstrapWhenNotified=y&:tabs=n&:toolbar=n&allowtransparency=true&Country=Afghanistan&:apiID=host0#navType=0&navSrc=Parse. Last accessed 2020/04/23
14. Institutional Development for Energy in Afghanistan (IDEA) Programme—GIZ: Enabling PV Afghanistan. Institutional Development for Energy in Afghanistan (IDEA) Programme—GIZ, Berlin, Germany (2017)
15. Ministry of Energy and Water (MEW)—Afghanistan: Afghanistan Renewable Energy Policy, Afghanistan Renewable Energy Policy (2017)
16. Renewable energy roadmap for Afghanistan (2015–2017). Asian Development Bank (ADB), Kabul, Afghanistan (2017)
17. World Energy Resources 2016. World Energy Council, London, UK (2016)
18. Mohammed, Y.S., Mustafa, M.W., Bashir, N.: Status of renewable energy consumption and developmental challenges in Sub-Sahara Africa. Renew. Sustain. Energy Rev. **27**, 453–463 (2013). https://doi.org/10.1016/j.rser.2013.06.044
19. Kariuki, D.: Barriers to renewable energy technologies development. https://www.energytoday.net/economics-policy/barriers-renewable-energy-technologies-development/. Last accessed 2019/11/01
20. Barriers to Technology Diffusion: The Case of Solar Thermal Technologies. https://webstore.iea.org/barriers-to-technology-diffusion-the-case-of-solar-thermal-technologies, (2006)
21. Radulovic, V.: Are new institutional economics enough? Promoting photovoltaics in India's agricultural sector. Energy Policy **33**, 1883–1899 (2005). https://doi.org/10.1016/j.enpol.2004.03.004

22. Huh, S.Y., Lee, C.Y.: Diffusion of renewable energy technologies in South Korea on incorporating their competitive interrelationships. Energy Policy **69**, 248–257 (2014). https://doi.org/10.1016/j.enpol.2014.02.028
23. Abdmouleh, Z., Alammari, R.A.M., Gastli, A.: Review of policies encouraging renewable energy integration & best practices. Renew. Sustain. Energy Rev. **45**, 249–262 (2015). https://doi.org/10.1016/j.rser.2015.01.035
24. Martinot, E.: Renewable energy investment by the World Bank. Energy Policy **29**, 689–699 (2001). https://doi.org/10.1016/S0301-4215(00)00151-8
25. Design and Performance of Policy Instruments to Promote the Development of Renewable Energy: Emerging Experience in Selected Developing Countries. The World Bank, Washington DC, USA (2012)
26. Public Utility Regulatory Policies Act of 1978 (PURPA) (1987). https://www.energy.gov/oe/services/electricity-policy-coordination-and-implementation/other-regulatory-efforts/public

Sustaining the Public Transport Network by Adaptation from Monocentric to Polycentric Structure

Naimatullah Shafaq Rahmatyar and Ujjal Chattaraj

1 Introduction

Nowadays, cities desires most to meet SDGs and their characteristics are applied practically in such a manner to support their sustainable development. However, cities are morphologically formed of two types, monocentric and polycentric. Each city may be functionally monocentric or polycentric with respect to their public transport network. This study focuses on Kabul city that is ranked the 5th fastest growing city of the world and has a morphological polycentric structure with monocentric public transport network. Such transportation network leads all main activities and trips to take place in the city center rather than other urban cores.

Here, we proposed a new integrated line linking the urban cores to each other using the headway-based assignment procedure of the *VISUM* macro simulation software. The new line, which is originated from Company Chowk (5th district of Kabul) and would be designated at Bot Khak Chowk (12th district of Kabul), pretends to strengthen the urban cores and create a more polycentric public transport network. Two variants were studied say *Variant* 1 and *Variant* 2. *Variant* 1 follows the present condition of the lines in Kabul city and its urban cores with buses with around 400 m distance between stops and the mean speed of 10 kph. *Variant* 2 follows the new planning for polycentrism, with the distance between stops around 800 m and a speed higher than *Variant* 1. The distance-based fare structure considered here was found to be the second key reason after the reduction of travel time for attracting people to use the new integrated line. Integrated line with optimal use of available buses

N. S. Rahmatyar (✉)
Baghlan University, Pol-e-Khomri 3601, Afghanistan
e-mail: nshafaq1989@gmail.com

U. Chattaraj
NIT Rourkela, Rourkela, Odisha 769008, India

M. S. S. Danish et al. (eds.), *Sustainability Outreach in Developing Countries*,
https://doi.org/10.1007/978-981-15-7179-4_9

eliminates the cars which are serving instead of public transport and contributes to sustainable development of the users, institutions, and environment.

Sustainable Development Goals (SDGs) are the most relevant targeted topic of different disciplines all over the world. The Afghanistan Sustainable Development Goals (A-SDGs) have been designed based on the council of minister's decision of October 7, 2015 [1]. In the meantime, the Afghanistan Green Urban Transport Strategy-AGUTS was officially released on November 2015 [2]. This research contributes to the 2nd characteristic "providing access to safe, affordable, accessible and sustainable transport for all" of the 11th goal "smart cities and communities" of A-SDGs. The demand for consuming the natural resources is more compared to their production; by knowing this fact, we have to optimally utilize the natural resources. The natural resource can easily be damaged due to our decreased attention for saving them from negative effects. These effects are mostly caused by humans due to rapid urbanization and less attention to the SDGs.

Up to early 1990s, the Trolley-Bus Public Corporation and Millie Bus Public Corporation used to operate more than 1000 vehicles for public transportation widely used by the citizens. The civil war, however, stagnated investment for route buses and trolley buses, and eventually the trolley buses failed due to destruction. Also less attention to the public buses along with lack of spare parts has reduced the operating number of buses gradually to only 105 buses as of August 2002 including 50 buses granted by the Government of India in August 2002. Nowadays the public transportation in Kabul city consists of small buses (Town ace and Mercedes), Large buses (public and private), and taxis. Bus services are operated by public bus sectors (Millie Bus) under the authorization of the Ministry of Transport and private enterprises.

The old buses which belong to Millie Bus are operating at a pace of only once an hour for each route which leads to the capacity declination of the public transportation. The disorder due to the conflicts which lasted over almost three decades has caused delays in road maintenance and stagnation of investment for the buses which are the major public transportation means.

A poor public transportation system and its low capacity paves the road for private buses and cars to operate more. Since the private cars and taxis with their low capacity (5 persons) contributed in transferring people, they have become the main cause of increasing traffic congestion, high fare, and air pollution [3]. To know better about cities, it is better to know about city models. The monocentric city model was developed from an approximation of Von Thunen, 1863 [4]. He described the relation between farmland values and accessibility. Later, this model was improved by others [5]. This model has been a very useful instrument due to its simplicity, exactitude, and capacity to be contrasted empirically. This model was assuming that all of the workplaces are located in the inner city and the families would compete for the houses with better access to the inner city [5].

The monocentric city model with its assumption has been criticized on the ground that the cities it explains are from a different era. It cannot explain the organization and configuration of the big cities. As they grow in size. The original monocentric structure of large metropolises tends to dissolve progressively into a polycentric structure over time. The central business district (CBD) loses its primacy, and clusters

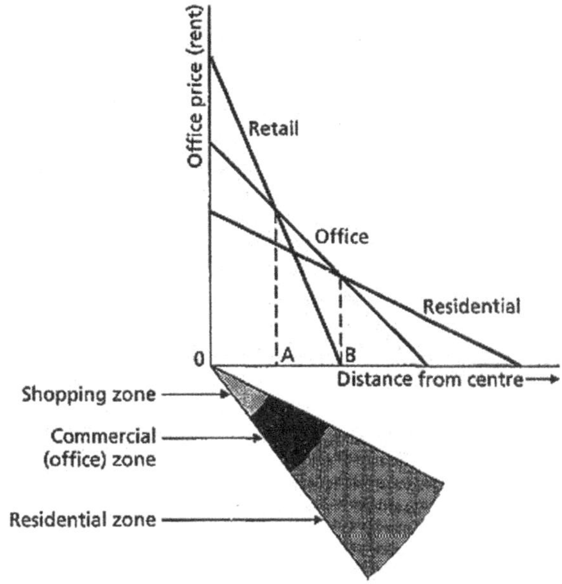

Fig. 1 Bid-rent theory [7]

of activities generating trips are spreading within the built-up area. The big cities are not formed polycentric at the beginnings. They may evolve in that direction. Monocentric and polycentric cities are animals from the same species observed at a different time during their evolutionary process [6]. Structure and functionality of a city are two different aspects. A city can be morphologically polycentric but not to work as a polycentric city [7] (Figs. 1 and 2).

From the morphological point of view, the cities are divided into two broad categories say morphologically monocentric and morphologically polycentric cities, where, from the transportation network point of view, they are divided into functionally monocentric and functionally polycentric cities with respect to their transportation network functionality [7]. Kabul, the capital city of Afghanistan, which is located in the northeast side of the country, with the total area of 1022.7 km^2 14 times that of 1964 consisting of 22 districts [3] with a projected population up to 6, 271, 710 inhabitants by 2025, is morphologically polycentric city [8].

Despite the fact that Kabul is the 5th rapid growing city of the world with increase in population from approximately 380,000 in 1964 to almost 4,825,900 inhabitants in 2016 [9], it is still functionally monocentric city. Nowadays the transportation network of Kabul city is monocentric, because all main lines are designed such that to originate/destinate into the city center and the urban cores are not connected to each other. Nowadays people living in Kabul are mostly suffering from the inadequate assignment of the public transport to connect the suburbs to each other and the city center. Even due to more congestion and less accessibility to the public transport the people are willing to either walk (for some kilometers) or to travel by car (with costs

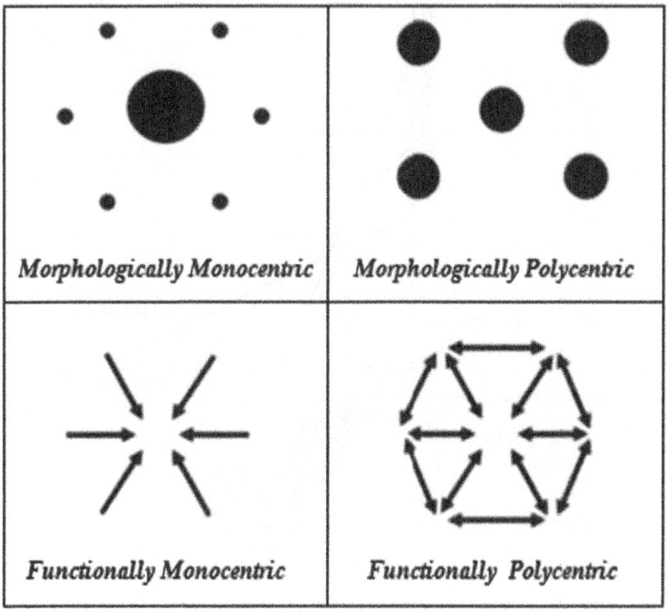

Fig. 2 Functional ploycentricity vs morphological ploycentricity [7]

more than four times that of bus fare). On the other hand, Millie Bus (National bus) has always been donated a huge number of buses [10].

Due to lack of such planning and management which could utilize the available buses, by which to serve the urban transport requirements, unfortunately the buses have been kept unused. Therefore, nowadays the private sector dominates almost totally the passenger transport sector in Afghanistan. Besides the agency considerations, the city structure also has a significant role in facilitating transportation services in the city. From the efficiency of source allocating point of view, monocentric city model often strongly depends on the portal function of the central city, while the overconcentrated public service function's monocentricity model weakens the competitiveness of secondary-size cities.

The polycentric network urban system had open internal as well as an external connection which can further enhance the externality of urban network avoiding the city size limitation to some extent and can, therefore, achieve complementary urban function in polycentric and networked mode [11]. A city with the morphologically polycentric structure having monocentric transportation network planning cannot satisfy the user demand with desired comfort level as well as the operator cost, because of causing more PJT, delay, congestion, and high fare. For such cities, the concept of functional polycentrism comes into mind. It is nothing but to transform the transportation network from monocentric to the polycentric structure, also integrating the existing modes of transport. Here, the public transport network plays the key role by connecting urban cores to each other as well as to the city center. An attempt was

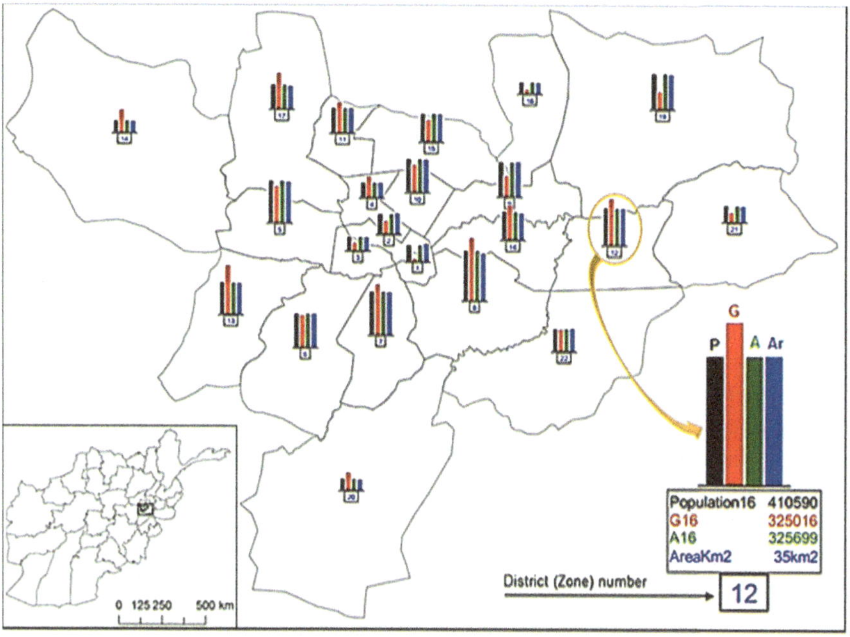

Fig. 3 Territory, socioeconomic chart, and location of Kabul city, 2016

made for integration of the public transport network using VISUM macro simulation software [12].

The methodology of applying the VISUM is stated as well. He also carried out the comparison between two variants to know which one to be considered as less or more integrated. Public transportation assignment consists of the three types of procedures, that is, Scheduled-based (SB), Headway-based (HB), and transportation system-based (TSB) assignment, among which the SB and HB assignments are more practiced in real life. To know in which condition the SB or HB have to be applied, findings of [13] show that for the higher frequencies the HB assignment behaves better, while for lower frequencies the SB assignment is better (Fig. 3).

2 The Proposed Solution Mechanism

- To understand what is the situation of the public transport network of Kabul city with respect to functional monocentrism and polycentrism;
- To know how to adapt the public transport network from functional monocentric to functional polycentric structure;

- To seek whether adapting the public transport network from functional monocentric to functional polycentric structure can help the sustainable development of the city's public transport;
- To propose a new line which can strengthen the polycentric structure of Kabul in a sustainable manner;
- To create, compare, and test different variants, to find the one which would help the sustainable operations of the network.

The deterioration of public transportation of Kabul city caused people to shift from a public mode of transportation to less efficient private cars which result in more environmental pollution and traffic jams. From the background study, we also learned about functional and morphological monocentricity and polycentricity. It was clearly observed that the morphological polycentric cities could be either functionally monocentric or polycentric with respect to their transportation network design to this fact. The main problem we considered here is the city which is morphologically polycentric but functionally monocentric.

Kabul is one of the cities of the world which is morphologically polycentric, but its transportation network is monocentric. All main activities and trips take place in the city center rather than other urban cores. Nowadays, more than 58% of the trips between the city's urban cores are made by passenger cars rather than buses, which is not cost-effective and does not support sustainability [3].

In this research, we made an attempt to see whether adapting the Kabul city public transport network from a monocentric to a functional polycentric structure can help to make it more functional, attractive, affordable, and accessible for users. For doing this, we considered the two main variants, Variant 1 and Variant 2. The solution mechanism itself consists of the following three major steps:

1. To create and modify the transportation network and to define the variants;
2. To apply the VISUM macro simulation model to each variant of the network;
3. To compare, analyze, and evaluate the variants.

Description of the 1st step of the above-mentioned step, which is nothing but network configuration and selection of variants, is provided in the below sections. The rest of the steps are clearly described in the methodology.

2.1 Network Configuration

A transportation network is typically referred to as a set of nodes, links, stations, lines, etc. In this study, the public transport network of Kabul was created in VISUM. Using the network editor menu of VISUM (Fig. 4), the public transport network was created.

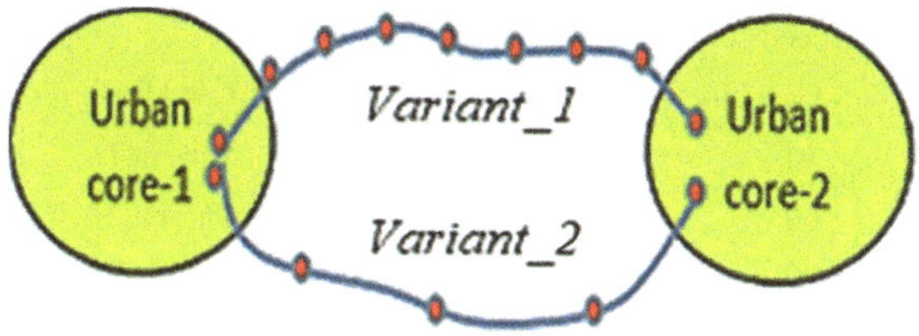

Fig. 4 Variants with different no. of stops

2.2 Selection of Variants

The main task here is to suggest different variants for the public transport network. We are going to find some optimal solutions for the improvement of the public transport network for today as well as for the future. From the background study, we understood that the present condition of the public transport network is too harmful to the people. That is why we are going to do some analysis for the improvement of the public transport network for today as well as the future. Although the solution for today's transportation network may not be practical, we want to show that using proper tools and methodologies, the weak transport networks can be analyzed for any given time which would lead to huge improvements with minimal cost and time. From the screen line survey, it was found that more daily trips within Kabul city are generated and attracted along the east–west axis rather than others so that we will study the route from Company Chowk to Bot Khak.

Also due to the existence of mountains in Kabul city and looking at the Green Urban Transport Strategy of Afghanistan (GUTSA), construction of new roads rather than improving the existing transport condition cannot solve today's problem because of land-use problem, tunnels required, and time taking to implement the big projects. Hence, first we try to overcome the problems that exist today by improving the available transport network rather than going for the construction of the new one. Then, the future planning for adapting from a monocentric to functional polycentric transportation network will be studied. Working with VISUM needs some characteristics to be introduced for the proposed line (new line) which are discussed later.

Distance between Stops: In the design of the line, by average, the distance between two successive stops should be fixed. Two main variants will be studied as follows:

- Line with more number of stops along the itinerary (present condition).
- Line with less number of stops along the itinerary (proposed condition).

Means of Transport: Typically in the public transport network, the distance between stops is bounded with the means of transport. So, the means of transport for each variant in 2016 and 2025 is given in the below lines, respectively.

1_Vehicles with low top speed will be the means of transport for Variant 1 (Present bus lines) for 2016.

2_ Vehicles with medium top speed will be the means of transport for Variant 1 (lines with large buses) for 2016.

Table 1 Capacity and speeds of different vehicles

Means of transport	Capacity (person)	Speed (kph)	Used for	Year
City current buses	30	10	Variant 1	2016 and 2025
Large buses	60	20	Variant 2	2016
BRT	90	30	Variant 2	2025

3_ Vehicles with low top speed will be the means of transport for Variant 1 (Present bus lines) for 2025.

4_ Vehicles with more top speed will be the means of transport for Variant 1. (BRT, LRT, City rail) for 2025.

Unfortunately, the city public transport is served only by bus lines and no more lines for BRT and LRT are open for operating. So, the means of transport for Variant 1 and Variant 2 in 2016 will be the present bus lines and the bus lines with large buses, respectively. Table 1 exhibits the capacities and speeds of different vehicles introduced to VISUM.

Variant 1: is taken as today's existing plan of the public transport network for Kabul for monocentrism consisting of bus lines with the distance between stops around 400 m. It is likely for people to take a bus with a reduced walk because of being more accessible for users regardless of the more journey time. From the O-D trips table that also shows the future growth of the trips, some of the urban cores were found to be located in districts 5, 2, 8, 12, 21, and 22 (New city of Kabul is not mentioned here). Furthermore, from the screen line survey conducted by [4], it is clear that the "Jadayi sehi Aqrab" road which connects districts 14, 5, 3, 13, 6, and 7 to district 2 (city center) carries the highest volume of traffic "49,800" PCU per 12 h in both directions during the period (7 am to 6 pm) and also the districts 21, 22, 18, and 19 in which the desired growth rate of generation is high compared to others. It is essential to study the lines L55 (Company to Mirwais maidan, L15 (Cinema Pamir to Mirwais Maidan) and L112 (Cinema Pamir to Bot Khak Chowk) and to continue the itinerary to connect district 21.

All the said lines are consisting of a few large buses, minibuses, microbuses, and taxis with poor services. The chosen values were given in Table 4.1. This variant consists of subvariants 1.1 and 1.2 stated below.

Variant 1.1: This Variant with total line route length of 30.394 km consists of the line routes L55, L15, and L112 and also the route from Bot Khak chowk (district 12) to Bot Khak (district 22).

Variant 1.2: This Variant with total line route length of 38.442 km consists of the line routes L55, L15, and L212 and also the route from Bot Khak chowk (district 12) to Bot Khak (district 22) (Figs. 5 and 6).

Variant 2: This variant consists of the same line routes as Variant 1, but with a reduced number of stops, an increased top speed of 20 and 30 kph for the years 2016 and 2025, respectively. It would be the planning which could simultaneously support the improvement of today's poor condition of public transport as well as adapting to

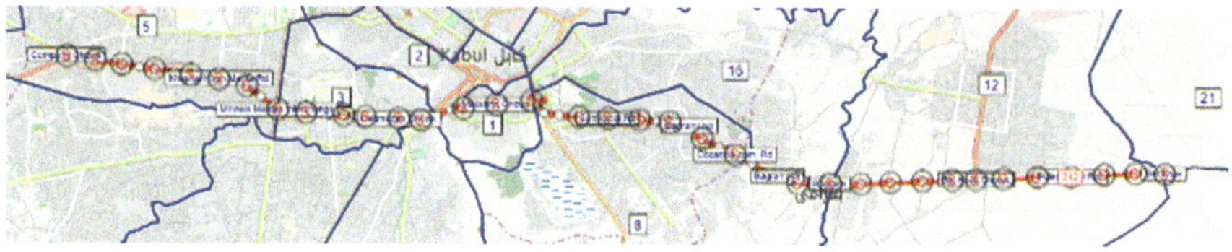

Fig. 5 Line route for Variant 1.1

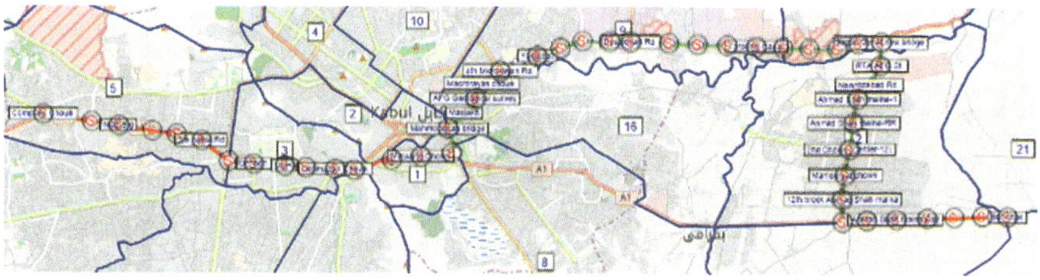

Fig. 6 Line route for Variant 1.2

the polycentrism in Kabul due to the reduction in travel time. It would be attractive to users. The stops meeting the following criteria are removed from each subvariant.

- 1- Stops with the sum of the minimum number of passengers boarding and alighting.
- 2- If the distance between successive stops on the route is > 800 m (Fig. 7).

3 Methodology

The research methodology framework, schematic of the four-step model running in VISUM, and VISUM input–output data are shown in Fig. 8. VISUM is the simulation software used for the present analysis. Using VISUM, the territory map of the Kabul city was created with proper division and numbering of the zones, nodes, links, and stops. User-defined attributes for each zone were introduced; later the centroid connectors for each zone were created to connect each zone centroid to the nearest nodes. Furthermore, the public transport network with existing line routes and stop points were established. Collected data were processed then introduced as attributes for each zone, node, link, and stop in VISUM [14].

From the Tsys/Demand/Dseg menu of VISUM, three transport modes, Public Transport (PuT), Private Transport (PrT), and Walking, were established; then, the time profile was set. The reference scenario that has to be introduced here is the "Do-nothing scenario." In this study, the only model to be used is the route choice model of the VISUM with the headway-based assignment (HBA) method. This model works with a static O-D matrix and network supply. Public Transport assignment procedures in VISUM use an impedance function to determine the impedance of a

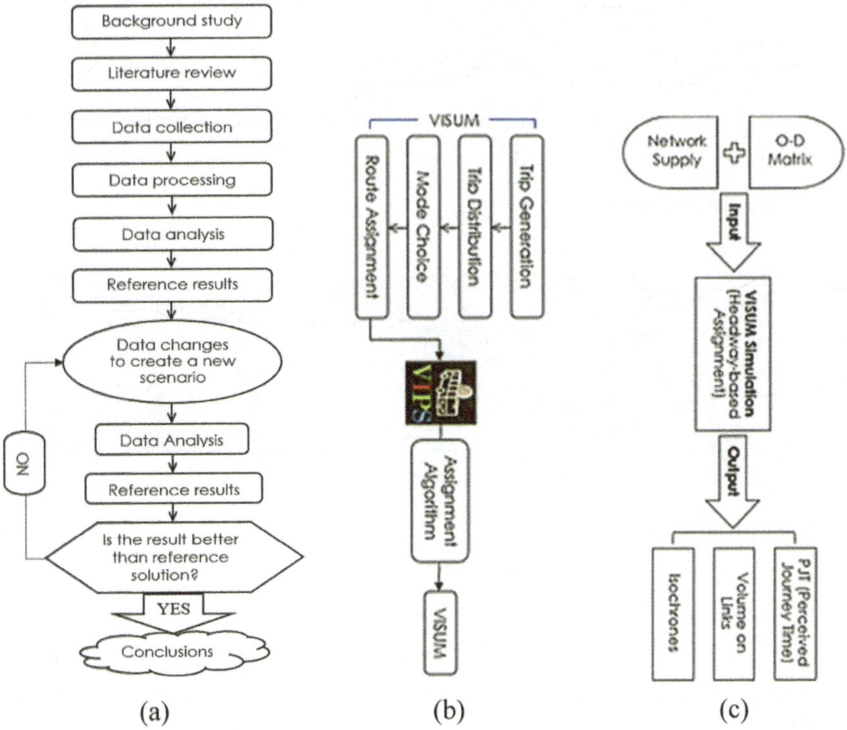

Fig. 7 **a** Methodology Framework. **b** Schematic of the four-step model running in VISUM. **c** The input–output data

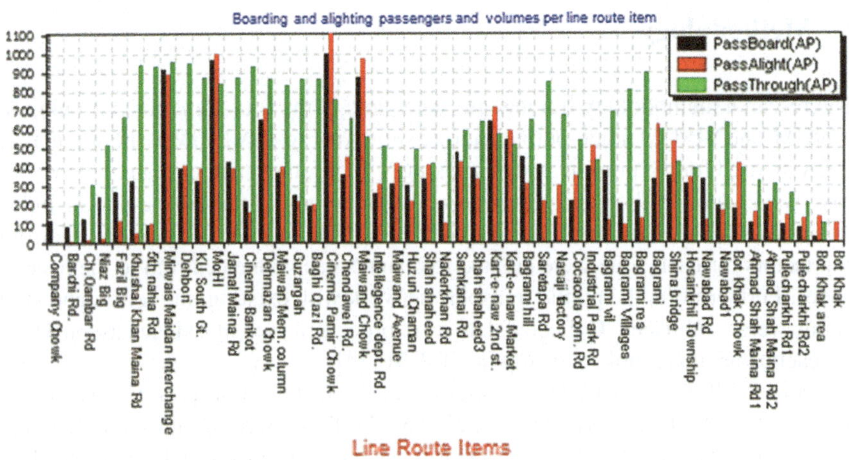

Fig. 8 Column chart for Variant 1.1

particular connection from several indicators of this connection [14]. In contrast to PrT, however, this impedance is used by some Public Transport procedures not only for the connection search but also to evaluate the connections prior to the connection split. Impedance can consist of time indicators and travel costs. The connection indicators can be divided into the following five categories: time indicators, length indicators, frequency indicators, monetary indicators, and derived indicators.

For the headway-based assignment, it is required to introduce impedance of the route given in Eq. 1 [14].

$$Impedance = Factor \times PJT + Factor \times Fare \tag{1}$$

VISUM calculates the headway-based assignment in order to minimize the expected travel time for all the demand. PJT consists of the weighted components of journey time, which affect the generalized journey time "minutes," shown in Eq. 2.

$$\begin{aligned} PJT_{.} = {} & w_{IVT} x\, t_{IVT} + w_{AUX} x\, t_{AUX} + w_{AT} x\, t_{AT} + w_{ET} x\, t_{ET} \\ & + w_{WT} x\, t_{WT} + w_{OWT} x\, t_{OWT} + w_{TWT} x\, t_{TWT} \\ & + w_{NT} x\, n_{TR} + BP_{PUT} + BP_{AUX} + P_{MD} \end{aligned} \tag{2}$$

where, F_1: Factor for PJT = 1.20, F_2: Factor for fare = 1.10.

W_{IVT}: weightage for in-vehicle time = 1.00, W_{AUX}: weightage for Put-Aux ride time = 1.00.

W_{AT}: weightage for access time = 1.80, W_{ET}: weightage for egress time = 1.80.

W_{WT}: weightage for walk time = 1.50, W_{OWT}: weightage for origin wait time = 1.80.

W_{TWT}: weightage for transfer wait time = 1.80, W_{NT}: weightage for number of transfers = 5 min.

t_{IVT}: in-vehicle time = (From simulation), t_{AUX}: Put-Aux ride time = (From simulation).

t_{AT}: access time = (From simulation), t_{ET}: egress time = (From simulation).

t_{WT}: walk time = (From simulation), t_{OWT}: origin wait time = (From simulation).

t_{TWT}: transfer wait time = (From simulation), n_{NT}: number of transfers = (From simulation).

BP_{PUT}: boarding penalty for PuT = 0, BP_{AUX}: boarding penalty for PuT-Aux = 0,

MD: mean delay = 0.

4 Result and Discussion

The results would be analyzed and compared for both the base year 2016 and for the target year 2025, to see what can help us in the improvement of the present as well as future conditions. The results will be analyzed by a comparison of the following three measures with the reference scenario then between each variant and subvariant to find the best one among all:

- Column Charts (Boarding, Alighting, and Through Passengers) on each particular stop.
- Total volume of passengers per each line route.
- Total Perceived Journey Times (PJTs) of passengers per zones.

Once all results are obtained, they would be compared, such that the Variant meeting less PJT, increase of the volume of passengers on each line route, and more passengers boarding, alighting, and going through each stop of each variant will be chosen as the best variant among all.

The variant found as the best one will be further studied to find the frequency of the line for different comfort levels. The following four comfort levels will be considered here:

1_Maximum capacity (seating 100% + standing 100%).
2_Bad comfort (seating 100% + standing 40%).
3_Fairly good comfort (seating 100% + standing 20%).
4_Good comfort (seating 100% + standing 0%).

Now we are going to compare the variants for the year 2016 and 2025.

4.1 Variant 1.1 Versus Reference Scenario (As Per 2016)

Column chart: Fig. 9 illustrates the column chart for direction "up." Column chart for direction "down" has the same structure as "up." From the column chart, the maximum number of passengers going through has increased from 598 to 962 persons compared to the reference scenario. It is a good change in planning. Important stops with high boarding and alighting rate are visible from the column chart as well.

Volume of passengers on each line route: The volume of passengers shows more improvement compared to reference scenario in the Huzori Chaman-Bagrami hill–Bot Khak Chowk; then Bot Khak as illustrated in Fig. 10. It means people are more interested in using this line compared to reference scenario (using four line routes from origin to destination).

Perceived journey time: The difference between perceived journey times for all the Variants is shown in Table 2.

4.2 Variant 1.2 Versus Variant 1.1 (As Per 2016)

Column chart: From column chart for the direction "up," it was seen that the maximum number of passengers going through has decreased from 962 to 911 persons compared to Variant 1.1. The column chart of direction "down" has the same structure as "up."

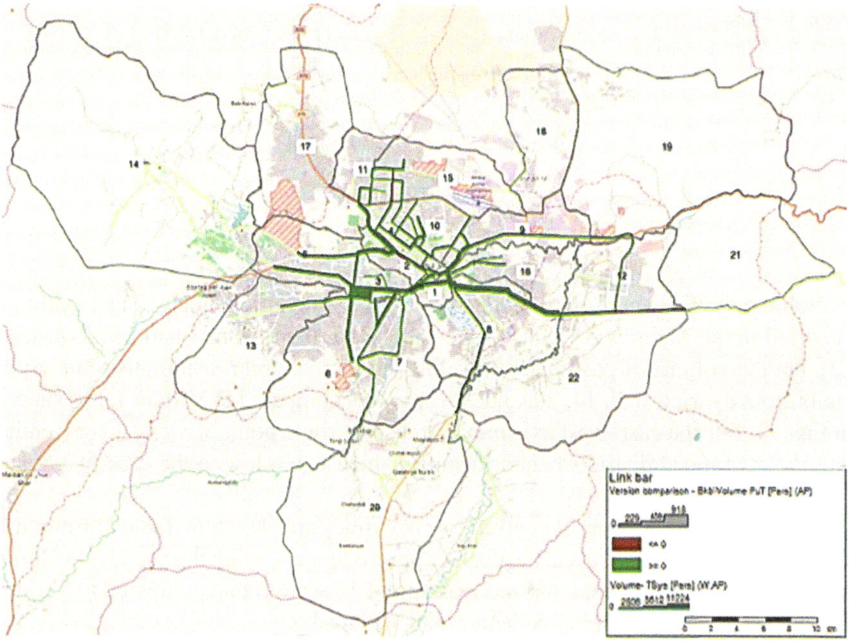

Fig. 9 Volume on each line route for Variant 1.1

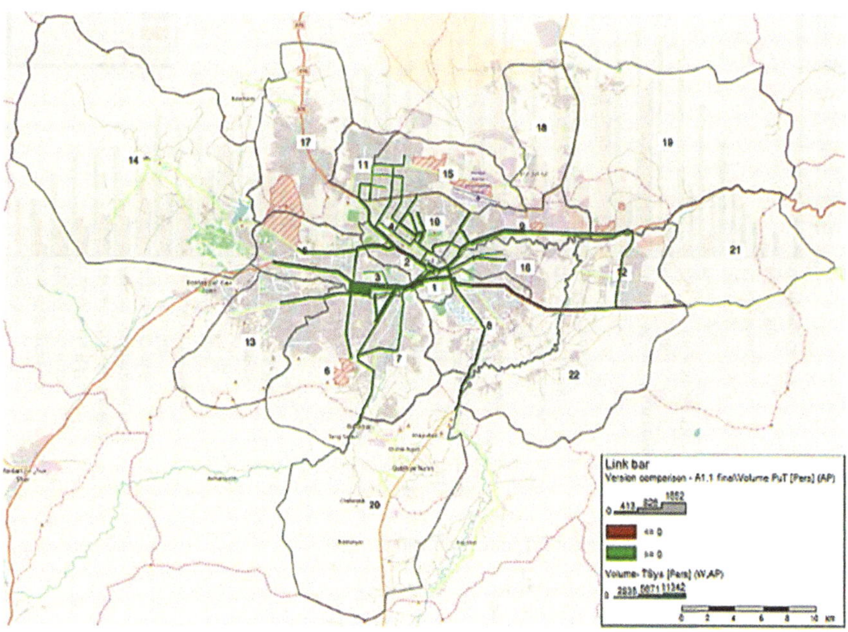

Fig. 10 Volume on each line route for Variant 1.2

Table 2 Comparison of TPJTs for all subvariants (2016)

Variants	TPJT of Alt-1.1—TPJT of Alt (i)
Reference scenario	−522
Variant 1.1	0
Variant 1.2	−124
Variant 2.1	327
Variant 2.2	−105

Volume of passengers on each line route: An improvement of passengers volume occurred in the Mahmood Khan Bridge-Pulchakhi, then Ahmadshah Mina (district 12), but the volume of passengers traveling from other zones using this route especially from districts 6, 7, 14, 20, and 22 is too low compared to Variant 1.1 as shown in Fig. 11. For the east–west axis public transport route point of view, it is essential that the mentioned districts benefit from it, which is too low in the case of Variant 1.2.

Perceived journey time: Perceived journey times are shown in Table 2. From the comparisons shown above and the decrease in passengers going through, less volume improvement in the desired districts and more PJT were found compared to Variant 1.1. Variant 1.1 was found to be better than Variant 1.2.

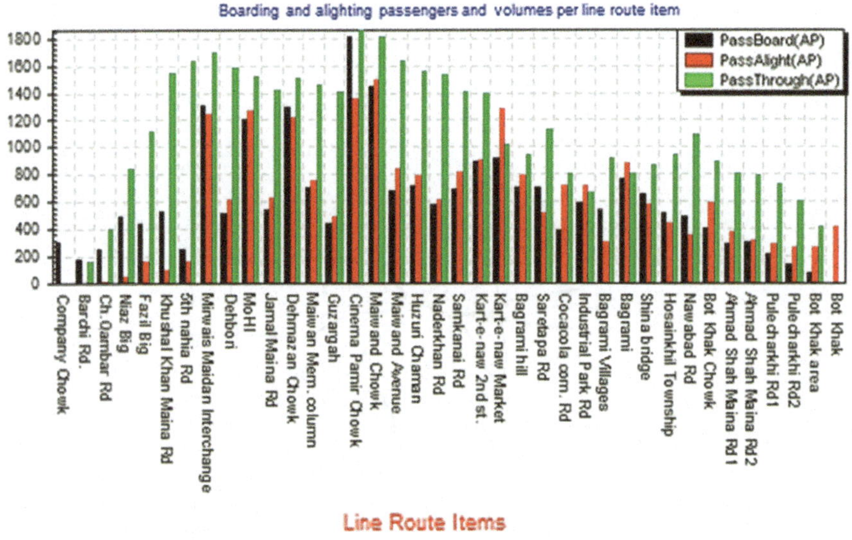

Fig. 11 Column chart for Variant 2.1, direction "up"

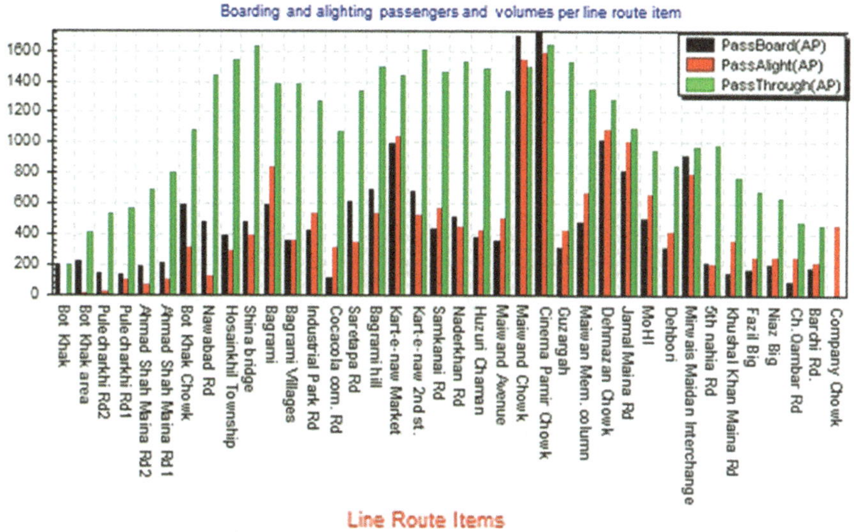

Fig. 12 Column chart for Variant 2.1, direction "down"

4.3 *Variant 2.1 Versus Variant 1.1 (As Per 2016)*

Variant 2 will be compared for the same line route with Variant 1. It means that we are going to know the effect of the new way of planning.

Column chart: For the direction "up" and the direction "down" the column charts can be seen from the below figures. The maximum number of passengers going through has increased from 962 to 1864 persons (almost twice). The main reason for this improvement is the high speed and accessibility of buses from any origin to any destination along the mentioned line. It is a very good change in this method of planning. Important stops with the high boarding and alighting passengers are visible from the column chart as well (Fig. 12).

Volume of passengers on each line route: Huge volume improvement occurred along the whole route especially from Bot Khak Chowk to Bot Khak which was always less in the case of other variants. Furthermore, a good volume improvement also occurred in the southern districts while average improvement is achieved in the north-eastern to north-western sides (Figs. 13 and 14).

Perceived journey time: Perceived journey times are shown in Table 2; Fig. 15.

4.4 *Variant 2.1 Versus Variant 1.2 (As Per 2016)*

Column chart: Column chart for the direction "up" can be seen from Fig. 16. The column chart of the direction "down" has the same structure as "up." The maximum

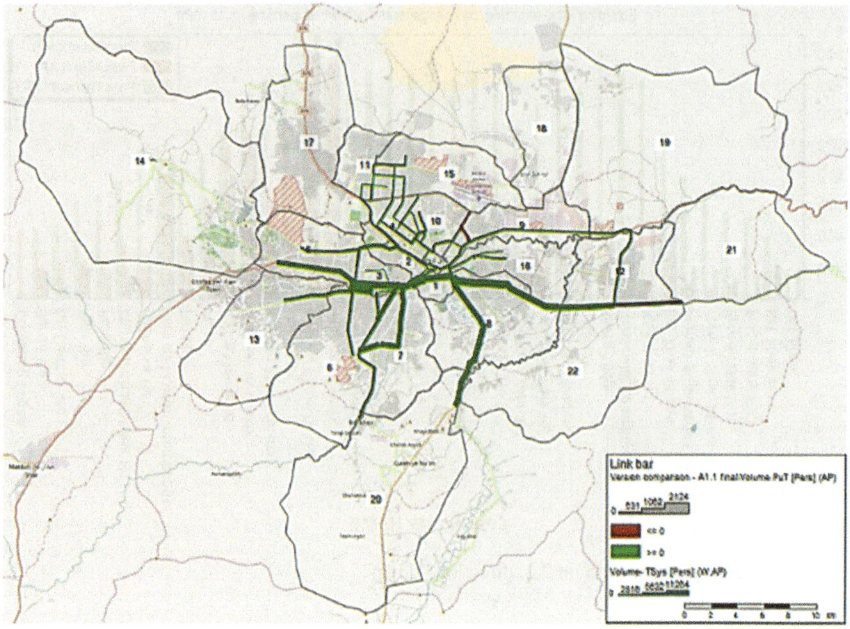

Fig. 13 Volume on each line route for Variant 2.1

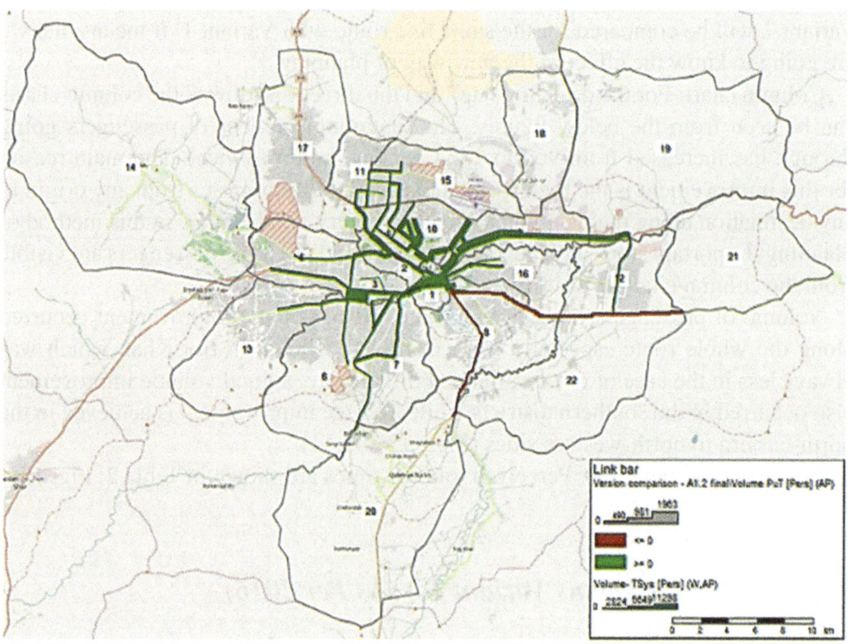

Fig. 14 Volume on each line route for Variant 2.2

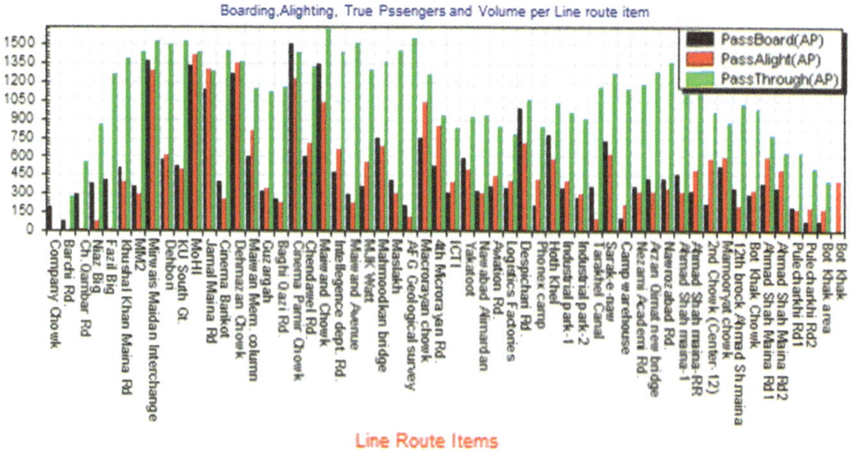

Fig. 15 Column chart for Variant 2.2, direction "up"

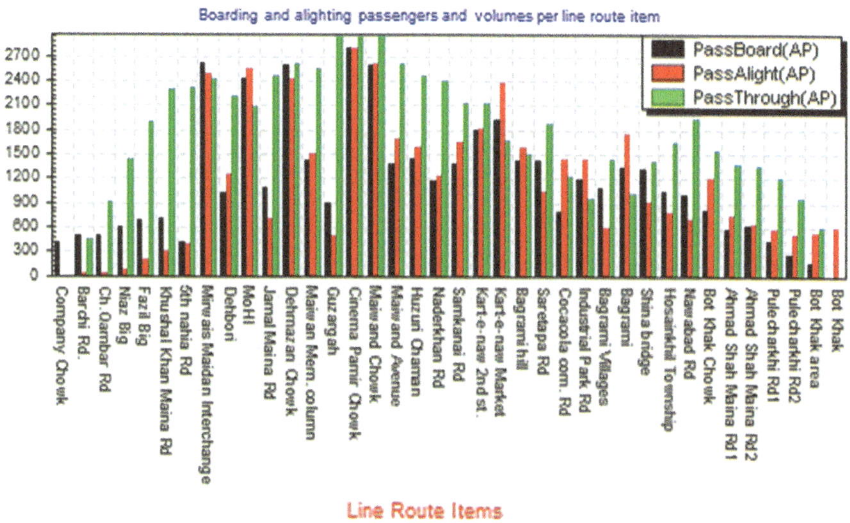

Fig. 16 Column chart for Variant 2.1, direction "up."

number of passengers going through has been increased from 911 to 1637 persons per four hours. It is very good change by Variant 2.2, but after comparison of volume and PJT, it will be decided whether this is the best of Variant 2.1.

Volume of passengers on each line route: Fig. 15 shows the volume on links for this subvariant. On the one hand, a good volume improvement occurred for the line route form Pulmahmood Khan to Pulcharkhi road and fair improvement in the Ahmadshah main area. On the other hand, this improvement is too low in other parts of the city and urban cores. It is not a good decision to adapt the network for that

Table 3 Number of passengers per hour for each vehicle type

		Vehicle type	Large bus (Millie bus)	Minibus (Coaster)	Microbus (Town Ace)	Taxi
Unit: Passenger		Capacity (Seating)	36	20	8	4
		Capacity (Standing)	24	5	0	0
		Total capacity (seat and standing)	60	25	8	4
Comfort levels	Headway (min)	Frequency (Veh/hr)	No. of passengers/hr			
Maximum Capacity	6	10	600	250	80	40
	5	12	720	300	96	48
Bad comfort	6	10	456	220	80	40
	5	12	547	264	96	48
Fairly good comfort	6	10	408	220	80	40
	5	12	490	252	96	48
Good comfort 6	6	10	360	200	80	40
	5	12	432	240	96	48

high demand of that particular area because the user cost will increase more which is not a good practice in transportation planning. The districts 6, 7, 8, 20, 21, and 22 show decrease in volume of passengers for such planning.

Perceived journey time: Perceived journey times are shown in Table 2. Table 3 shows the number of buses calculated for different comfort levels for the year 2016. Keeping the capacity of the BRTS bus as 90 passengers, using the same method, the frequency of line, the number of basses for different comfort levels can be calculated for the year 2025.

4.5 Variant 1.1 Versus Reference Scenario (As Per 2025)

Column chart: From the column chart, the maximum number of passengers going through has increased from 598 to 1928 persons compared to the reference scenario. Which is a good change in planning.

Volume of passengers on each line route: The volume of passengers shows more improvement compared to the reference scenario in the Kart-e-naw-Bot Khak. It also shows a small decrease on the Pule Charkhi road. It indicates the attractiveness of the line route.

Table 4 Comparison of TPJTs for all subvariants (2025)

Variants	TPJT of Variant (i)—TPJT of Variant 1.1
Reference scenario	−6957
Variant 1.1	0
Variant 1.2	−1423
Variant 2.1	***674***

Perceived journey time: The difference between perceived journey times for all the variants is shown in Table 4.

4.6 *Variant 1.2 Versus Variant 1.1 (As Per 2025)*

Column chart: Here, the maximum number of passengers going through has decreased from 1928 to 1787 persons compared to Variant 1.1.

Volume of passengers on each line route: Volume of passengers has an improvement on the Pule Charkhi and Ahmad Shah Mina road, but more decrease on the Kart-e-naw road as well as the routes to districts 7, 8, and 22 compared to Variant 1.1. It shows the relative deduction in volume, while the target route remains unserved.

Perceived journey time: The difference between perceived journey times for all the variants is provided in Table 4.

4.7 *Variant 2.1 Versus Variant 1.1 (As Per 2025)*

Column chart: A huge increase in the maximum number of passengers going through has occurred in this comparison. Compared to Variant 1.1, it shows an increase from 1928 to 2958 persons. It is a really great change in volume and indicates the proper attractiveness of the planning in this way. A column chart is shown in Fig. 16.

Volume of passengers on each line route: The volume of passengers has an improvement on the Pule Charkhi and Ahmad Shah Mina road, but more decrease on the Kart-e-naw road as well as the routes to districts 7, 8, and 22 compared to the Variant 1.1. It shows the relative deduction in volume, while the target route remains unserved.

Perceived journey time: The difference between perceived journey times for all the variants is provided in Table 4.

4.8 *Variant 2.2 Versus Variant 1.2 (As Per 2025)*

Column chart: Here, the maximum number of passengers going through has increased from 2351 to 1787 persons compared to Variant 1.1.

Volume of passengers on each line route: Volume of passengers has further improved on the Pule Charkhi and Ahmad Shah Mina road.

Perceived journey time: Table 4 provides the difference between total perceived journey times for all the variants.

5 Contributions and Findings

New finding to this research can be listed as follows:

- Monocentricity and polycentricity have different meanings with respect to the morphology of the cities and functionality of their transportation networks, because a city may be morphologically polycentric, but its transportation network may not work polycentric. So, such a city is called "morphologically polycentric but functionally monocentric."
- Integrating the present multiple line routes helps the sustainable development of the city such that, to reduce the traffic jam and air pollution and improved speed;
- Adaptation of public transport network from functionally monocentric structure to the functionally polycentric structure was examined. The procedures that need to be followed for the said purpose were clearly mentioned, which can be used by different researchers for different cities all over the world.
- An attempt was made to check a new integrated line of the Kabul city public transport network which would help adaptation from functionally monocentric to functionally polycentric structure. It was found that the mentioned adaptation is possible and also helpful, only by the manner to link each urban core among themselves as well as the CBD.
- The distance-based fare structure is more attractive and affordable for users compared to fixed fare structure.

6 Conclusion

An attempt was made in order to study what should be the best way to improve the public transport network in the suburbs, making the multiple-core structure more functional and stronger in a sustainable manner. The city is growing by around 100,000 people per year and it makes it necessary to adapt the public transport network to this fact in order to reduce the taxies and private cars and to get a more cohesive city with strong connections between the urban cores. Kabul city master plan and Public Transport Program of Kabul City also focus on this task, trying to plan suburbs that are more integrated with reinforced connectors and to further develop a dense, multicenter city. Thus, Kabul needs to develop a good public transport system to support a polycentric and densely populated structure. In this paper, both the political and technical interests are taken into consideration.

Kabul is more polycentric, but this does not only mean that there are some differences in the density pattern (there are many urban cores). It is needed that these cores represent real unifying elements of urban subsystems. The city should be planned in the future considering this situation. Urban cores of Kabul have almost the same structure and need to be developed in different ways to attract more people from other suburbs. This development will avoid travels to the city center that will help reducing congestion problems that currently exist. In the short term, and according to Kabul Urban Design Framework (KUDF) and Capital Region Independent Development Authority (CRIDA), the outer cores need to become more metropolitan in character. In addition, improving the accessibility is a prerequisite for this development in the regional cores. An integrated line like that is planned here will help to strengthen this connection between the inner parts as well as outer parts of Kabul in a sustainable manner. After the comparison of the two main variants, it was found that planning for polycentrism has more advantages than the monocentrism (the plan that exists today). Travel times were reduced which led the line to attract more people (Technical sustainability).

Also, the number of users would be more in the future as stated earlier. This program does not take into consideration people that could change their travel habits and would choose traveling by public transport instead of cars if the new line is attractive enough (environmental sustainability). Variant 1 gives more accessibility to the people who live between urban cores because it has more stops, but the travel time increases, and it does not help the polycentric concept. Therefore, this variant does not support polycentrism and sustainability. A vehicle with higher speed is really needed if the connections between urban cores needs to be improved.

Variant 2.1 gets more passengers crossing the northeast-southwest axis of Kabul than the other variants and the amount of travelers between north–south axis is well-balanced (Social sustainability). It also increases travelers among suburbs and decreases the number of people that travel to the city center, which helps avoiding congestion in the inner city (environmental & technical sustainability). Perceived journey time is reduced quite enough compared to the reference scenario. At the same time, the costs decreased because the new line is an integrated bus line for 2016 and BRTS line for 2025, which follows the route of other buses and private cars that already use these routes in a modified manner (economic & technical sustainability).

To maintain public transportation services in sustainable manner, it is necessary to establish distance-based fare structure and to manage and maintain the level of services in such a way that would make them competitive that leads to economic and social sustainability. This, in turn, will necessitate the management of routes, terminals, and maintenance with profitability in mind, which leads to institution sustainability. The problem with this variant compared to Variant 1 is that people who live between cores have fewer facilities to use this line. Thus, to make this line working better, it is necessary to have some lines that work as feeders to the main proposed line. Modifying the Kabul public transport network to functional polycentric structure contributes to the sustainable development of the city because the available infrastructure would be optimally utilized; also, the small line routes would be replaced with an integrated line. The increase in the number of buses and cars especially in the city center is expected to exacerbate air pollution.

To prevent this, better maintenance for bus and automobile transportation needs to be provided. Hence, the new integrated line would lead to a reduction of cars by replacing the high capacity buses, higher speed, and more distance between the stops. The passengers are not desired to pay the fixed fare. They have to pay a distance-based fare consisting of the boarding penalty along with the fare for traveled distance. In other words, the new line helps a person to travel fast and to pay less, which leads to meet technical, economic, environment, and social sustainability. The increased fleet size (the cars are desired to be replaced with buses), the integrated line, higher speed, more distance between stops and more attraction of people and minimal demand for new constructions would obviously lead to operational sustainability of the city's relevant authorities.

Finally, the proposed solution is environment friendly at all. To make these happen, it is strongly recommended to apply schedules for each line route, E-ticketing system, good maintenance to the road and buses, improving the infrastructures, and enforcement of traffic law. Depending on demands after pm peak hour of each line routes, it is strongly recommended to put some 2–3 buses on 1–2 h duty to take the passengers into their homes.

References

1. Sustainable Development Goal 12: https://sdgs.gov.af/109/sustainable-development-goal-12. Last Accessed 01 Nov 2019
2. Habibzai, A.J.: Afghanistan Green Urban Transport Strategy: 2015–2025, https://www.afghan engineers.org/wp-content/uploads/2017/01/Afghanistan-Green-Urban-Transport-Strategy-Abdullah-Habibzai.pdf (2015)
3. Draft Kabul City Master Plan: Product of Technical Cooperation Project for Promotion of Kabul Metropolitan Area Development Sub Project for Revise the Kabul City Master Plan, https://openjicareport.jica.go.jp/pdf/12058566_01.pdf, (2011)
4. Arnott, R.J., MeMillen, D.P. (eds.): A Companion to Urban Economics. WB Publishing Ltd., Massachusetts, USA (2007)
5. Alonso, W.: Location and Land Use: Toward a General Theory of Land Rent. Harward University Press, Massachusetts, USA (1964)
6. Bertaud, A.: The Spatial Organization of Cities: Deliberate Outcome or Unforeseen Consequence? Institute of Urban and Regional Development, University of California at Berkeley. 1–32 (2004)
7. Burger, M., Meijers, E.: Form Follows Function? Linking Morphological Funct. Polycentricity: Urban Stud.. **49**, 1127–1149 (2011). https://doi.org/10.1177/0042098011407095
8. Country Population Estimation: https://www.nsia.gov.af:8080/wp-content/uploads/2019/04/Final-Population-1396.pdf, (2016)
9. The world's fastest growing cities and urban areas from 2006 to 2020, https://www.citymayors.com/statistics/urban_growth1.html, Last Accessed 01 Nov 2019

10. Policy Paper 1.1: Corporatisation of MOT Truck and Bus Operations, https://siteresources.worldbank.org/EXTSARREGTOPTRANSPORT/Resources/579597-1128434742437/1735263-1128436052415/30MOT.pdf (2004)
11. Zhao, M., Chen, C.: Polycentric network organization of mega-city regions in yangtze river delta. Procedia Earth Planet. Sci. **2**, 309–314 (2011). https://doi.org/10.1016/j.proeps.2011.09.048
12. Solecka, K., Zak, J.: Integration of the Urban public transportation system with the application of traffic simulation. Transp. Res. Procedia **3**, 259–268 (2014). https://doi.org/10.1016/j.trpro.2014.10.005
13. Cascetta, E., Coppola, P.: Assessment of schedule-based and frequency-based assignment models for strategic and operational planning of high-speed rail services. Transp. Res. Part A: Policy Pract. **84**, 93–108 (2016). https://doi.org/10.1016/j.tra.2015.09.010
14. Group, P.: VISUM 12.5 Fundamentals. Epubli, Berlin, Germany (2012)

Sustainable Transportation and Mobility System in Kabul City

Homaira Mansoor, Nazifa Rasoli, Kh Jamilurahman Habibizada, Bashir Ahmad Raqi, Najib Rahman Sabory, and Ghulam Farooq Mansoor

1 Introduction

Transportation systems help the flow of materials and services around the world, creating jobs for many people and facilitates economic growth. Meanwhile, it generates negative environmental impacts on societies. Gas emission and consequent air pollution are some of the adverse effects of transportation systems [1]. Every person in the community is responsible for future generations to leave them with a livable planet that is their inalienable right. So there is a severe need to combat climate change and air pollution through ensuring sustainable transportation mechanisms. Sustainable transportation is defined as "satisfying present transport and mobility needs without compromising the ability of future generations to meet their needs" [2]. Transportation is the leading cause of air pollution in Afghanistan. There is no accurate data from gas emissions contributed by transport sector in Afghanistan. In Europe and rest of the globe, transport sector stands for about a quarter of the greenhouse gas emissions [3]. As sustainability is an integrated concept of business activities and includes social, environmental, and economic dimensions, hence ensuring sustainability has health, economy, society, and environmental benefits [4]. Reducing air pollution affects human activities, future productivity, income, and economic growth [5, 6].

Furthermore, diseases caused by air pollution have expensive medical costs. Afghanistan being a developing country, the choice of the fuel type is strongly influenced by the relative price. Importantly, transportation plays a prominent role in SDGs ensuring. It directly contributes to good health and wellbeing in all ages, food

H. Mansoor (✉) · N. Rasoli · K. J. Habibizada · B. A. Raqi · N. R. Sabory
Kabul University, Kabul 1006, Afghanistan
e-mail: homairaman@gmail.com

G. F. Mansoor
Health, Nutrition and Social Science Research, Freelance Researcher, Kabul, Afghanistan

M. S. S. Danish et al. (eds.), *Sustainability Outreach in Developing Countries*,
https://doi.org/10.1007/978-981-15-7179-4_10

security, affordable and clean energy, and combating climate change. It can indirectly reduce poverty by making transportation efficient and affordable [7].

An effective transportation system should consider accessibility and affordability by all users, travel time reliability, safety, security, cleanliness, and multimodality. In the long term, use of smart technology can enhance efficiency of mobility and transportation.

Decades of war in Afghanistan destroyed overall infrastructures including roads and passways in main cities such as Kabul. Population of Kabul city has increased dramatically after the civil war due to the returning of refugees from Pakistan and Iran, and displacement of people from different provinces. Moving people and goods is an extreme need from the past till now. The city faces a lack of managed transport infrastructure and access to basic services. On the other hand, geographical position and topography of Kabul is a challenge beyond transportation and mobility.

The Afghan government, with support from donor communities started to rebuild streets and roads. However, full consideration of the looming growth and influx of people from different countries and provinces are ignored. An effective public bus system was built in 1976, which is destroyed during the post-war. Besides, the quality of the reconstructed roads was so poor, most of the streets were destroyed after a few years of rehabilitation/reconstruction. The existing streets are insufficient to meet the citizens' needs and to achieve rapid development and economic growth. This poorly managed transportation and mobility have consequences for health, environment, and economy. However, there has been no systematic assessment to explore roots of these problems and evaluate their impact on the environment, health, and economy.

2 Proposed Solution Mechanism

During June-July 2019, an assessment is conducted to explore mobility issues in Kabul city and to figure out municipality response to mobility needs of the growing population of Kabul. Specific objectives of the assessment are as follows.

- To assess the extent and scope of the problem of transportation, streets, and infrastructure.
- To explore the effects of the current situation of transportation on the environment, health, economy, society, and security.
- To make recommendations for improved mobility and transportation in Kabul city.

In order to address the objectives, we used a cross-sectional design and mixed methods. Mixed methods included visits and observation of crowded intersection, pedestrian, and underground bridges, besides questionnaires with Kabul residents and face to face interviews with key informants. Secondary data about vehicles of Kabul city obtained from the traffic management office in Kabul. Information about Kabul's transportation infrastructure was obtained from Kabul municipality and the infrastructure office. Literature review included peer-reviewed articles and available government documents.

Sampling strategy: both existing underground pedestrian bridges were observed. Of the flyovers, only one was observed as all had similar characteristics of design, location, and challenges. For face to face interviews, key departments that were believed to be able to provide data on mobility issues were targeted. Finally, based on the availability of resources, a total of 40 residents of Kabul were selected randomly at the center of Kabul city near Shahe Dushamshera for the application of the questionnaire.

Ethical considerations during collecting data: before starting the interview, study objectives and processes involved and risks and benefits were explained. Potential participants were told that participation was voluntary and assured that information they provide would be kept confidential, and their identity will be anonymous.

Analysis: quantitative data were analyzed descriptively. Frequency and percentages of key indicators were produced, then they are visualized using a pie chart. Qualitative data, such as notes from the observation and photographs, as well as qualitative interviews, were applied to key themes. Photos were used in the report to elaborate on the situation further.

3 Results and Discussion

A total of 40 residents from Kabul city center participated in the quantitative interviews. Qualitative interviews were conducted with the head of traffic, Kabul Municipality mayor, and the head of the infrastructure at the Ministry of Urban Development. Findings of the literature review and field data collection focus on the situation of transportation in Kabul and the effects of the situation on health, environment, society, economy, and security. Current status of transportation in Kabul is presented in the following scenarios:

- Existing transportation modes in Kabul.
- Current status of transportation infrastructures.

3.1 *Kabul City Existing Transportation Modes*

Due to rapid population growth, Kabul city faced numerous challenges and constraints with an extreme need for public transportation. However, existing public transportation is not able to meet demands. Passengers always overload a few available public buses. The busses carry many people where many passengers stand in the aisle of buses. Number of private vehicles in Kabul city is getting higher each year due to population growth, lack of public transportation, lower price, and lack of regulation on controlling importing second-hand vehicles.

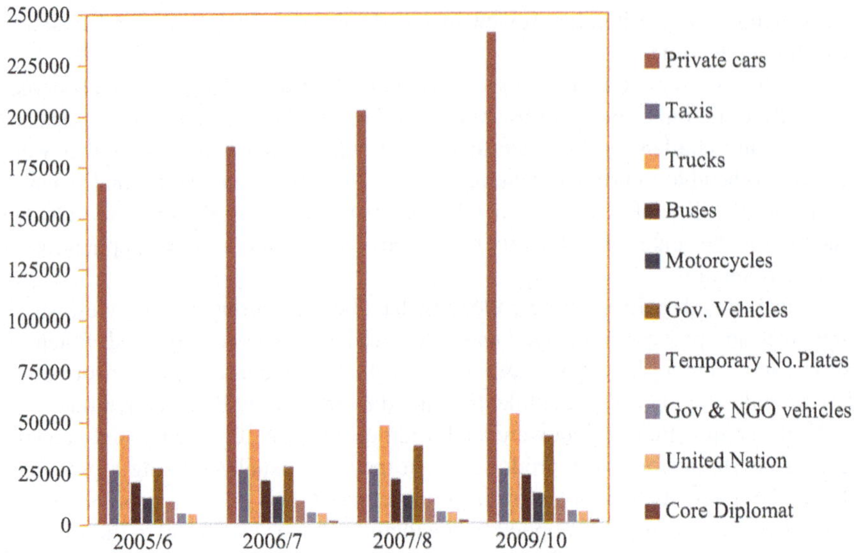

Fig. 1 Number of cars in Kabul from 2005 to 2010 [8]

Private unlicensed taxis are getting higher in numbers due to high price of the license of taxis. Taxi share system is cheaper as it has a fixed rate of 20 Afg (approx. US $ 0.25) per person no matter what distance is in the city. Motorcycle and bicycle are used as mobility means and also carry goods. But due to cultural issues, women are not allowed to ride a motorcycle or a bicycle. Also, air pollution prevents people from riding a bicycle and motorcycle. Figure 1 shows number of cars in Kabul city from 2005 to 2010. Number of private cars has increased at an unprecedented rate relative to other modes of transportation.

3.2 *The Current Situation of Transportation Infrastructures*

Technical transportation infrastructure includes roads, railways, bridges, tunnels, ports, airports, urban transportation infrastructure, dry ports, inland container depots, pipelines, and signage, and traffic management systems [9]. In the case of urban areas, technical urban transportation infrastructure includes roads, tramways, bus lanes, terminals, bus stations, and bus stops, parking places, bicycle ways, footpaths, pedestrian crossings (zebra cross, footbridges, and underground passes), road junctions (intersections and interchanges), bridges, tunnels and signage, and traffic management systems (electronic devices) [10].

Roads and streets are the only main transportation infrastructure in Kabul city. Kabul's roads are in poor condition, and the municipality has no sufficient budget to construct new infrastructure and maintain the existing infrastructure. Overcrowded areas in Kabul city, Deh-Mazang, Jada-e-Maiwand (Fig. 2) Baghbala, Sarae-Shamali, Karte-Parwan, Karte-Maureen, Kote-Sangi, Poli-Sokhta, and Char-Qala, face the most severe traffic congestion.

The existing road network has 3 types of roads (Fig. 3): arterial roads that are wide and paved, secondary roads that connect to arterial roads, and have 2–3 paved lanes and neighborhood roads that are mostly unpaved and have 2 lanes. According

Fig. 2 An overcrowded area in Kabul city (Jada-e-Maiwand) [11]

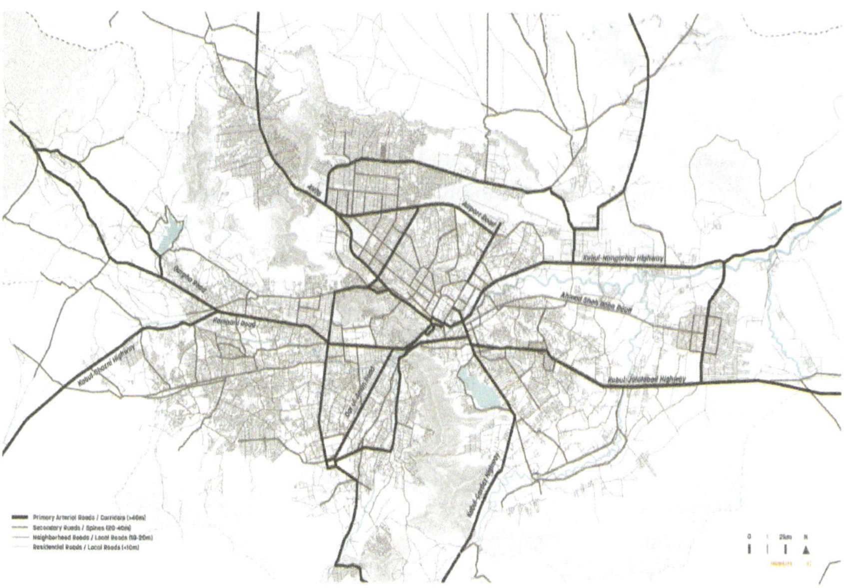

Fig. 3 The existing road network of Kabul city [12]

Fig. 4 View of the crowded street by hawkers and cars in Kabul city [13]

to the Kabul municipality development program, almost all arterial and community streets are paved and upgraded, and many secondary and neighborhood roads are being paved each year.

Streets in Kabul city are heavily crowded due to the blocking of many roads for security issues. Two sides of the road are occupied by vandals and hawkers, which causes a lack of parking space for parking private and public transportation, especially in bazaars and market places in Kabul city (Fig. 4). In unplanned areas, there is no paved road for vehicles.

Roads can be used only by a pedestrian or only one car can pass. During the winter, these roads are muddy and hard to use by pedestrians and vehicles.

Footpaths are occupied by vandals and hawkers and blocked by some concrete walls that prevent pedestrians from using footpaths. So pedestrians use the streets to walk, which disorders the traffic flow and creates huge congestion. To solve the traffic congestion, which is made by pedestrians in crowded intersections, the municipality made many bridges and underground paths for pedestrians to pass over or under intersections without interrupting the traffic. But both were not successful, and people still use the streets.

As the number of vehicles in Kabul increases, demand for roadside parking and specific parking places increases. At present, Kabul city faces a lack of parking areas for private and public vehicles. Whereas roadside parking is the main issue for the flow of traffic. City lacks well-designed bus stops and transit stops that buses and taxis stop and collect passengers without disturbing the traffic flow.

Little working on signage and traffic lighting in Kabul city has been done. Many intersections are controlled by traffic lighting, but still, people do not respect the traffic rules, and many other intersections are controlled by two or three traffic policemen. Most of the intersections are neither controlled by traffic lighting nor by traffic policemen. Weak traffic management caused negative impacts such as air pollution, fuel consumption, travel time, occupancy of the park, an increase of accidents, traffic congestion, and an increase in distance to destination. Other challenges in Kabul city can be listed as follows:

- Lack of sufficient funds for new construction and maintenance of the transportation infrastructures.
- Low awareness among people about use and maintenance of transportation infrastructures.
- Most activities and services are centralized in one area, which causes congestion.
- There is little or no supervision of regulations on importing fuel in Kabul city.

3.3 Effects of the Current Situation of Transportation in Kabul City

Poor transportation system in Kabul has affected environment, health, society, economy, and security. Kabul city cannot handle more than 70,000 vehicles, while there are more than 400,000 vehicles in Kabul city [14]. Below we explain each one of them.

Environmental Impact: Vehicles are one of the main air pollution sources in Kabul city. Cars and vehicles' pollutants have long terms effects on environment as well. The most important impacts of transportation on environment are as following.

- Climate change: Atmosphere consists of greenhouse gasses that absorb or emit the radiation of the sun; thus, it keeps our earth warm. But excessive amounts of greenhouse gasses ruin the balance between absorbing and emitting sun radiation, which contributes to climate change. Vehicles are the major cause of climate change that emit main greenhouse gasses such as carbon dioxide, nitrous oxide, and methane.
- Abolishing valuable agricultural lands: urban expansion and transportation infrastructures cause abolishing valuable agricultural lands and habitat fragmentation.
- Water and soil quality: Due to carbon dioxide gas emission from vehicles, rainwater becomes polluted and acidic. Using toxic material by vehicles lead to soil contamination.

The effects of the current situation of transportation on health (Fig. 5) are as follows:

- Effects on the respiratory tract: Children are generally more vulnerable to the hazards of air pollution than adults. Children's lungs are in progress of development and are prone to many lung diseases such as bronchitis and pneumonia.
- Effects on cognitive development: Researches show that air pollution also contributes to damaging cognitive development in children and fetuses. Over the past decade, researchers have found that high levels of air pollution may damage children's cognitive abilities, increase adults' risk of cognitive decline,

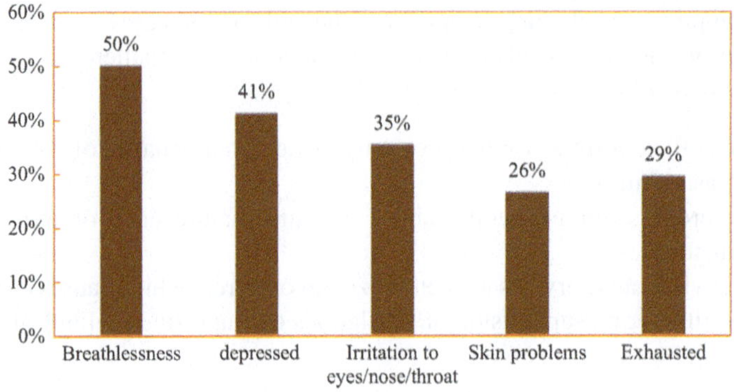

Fig. 5 Participants' perception of how air pollution affected them

and possibly even contribute to depression [15]. Studies have shown that "because these particles enter the bloodstream, they can also cross the placental barrier and affect the development of the fetus, including both physical and cognitive development" [16, 17]. It can cause future developmental delays and lower verbal intelligence quotient (IQ) [18]. Consequently, it has lifelong implications for school and future career and totally affects the country economy.

- Immune system: The immune system of children who are exposed to air pollution will be compromised and lose its function to combat viruses, and their bodies will get weak toward viruses and bacteria.
- Fetal loss and premature delivery: After children, women are more vulnerable to air pollution, which harms their reproductive cycle, and for pregnant women, there is a risk like fetal loss and premature delivery.

3.4 Effects of the Current Situation of Transportation on Society

Transportation plays an important role in society, and it has direct effects on the quality of life and improving societies. It enables people to meet and connect to the community, and it provides access to services for people.

In Kabul city, existing and new transportation infrastructure is not people-centered. It is not equally accessible to all users, while disabled people, elder, and low-income people's demands are not considered. Preserving architectural heritages and buildings have high value in Afghanistan. As human's health is affected by pollution, buildings and heritages are also damaged by air pollution. Most of Kabul heritage is located in crowded areas where smoke from most old-tech cars engines deposits on the building's facade and reduces the service life of materials used in building. Also, the aesthetic purpose of the painted facade of heritages is lost due to dust. The dirty facade has a negative effect on pedestrians, and it is not pleasant for pedestrians to use pathways.

Economic impact: Poor quality of transportation can limit many trades and can treat economy. Creation of a green transportation system in Kabul city has great value.

Transport systems in Kabul city are deficient in terms of capacity and time reliability. Transferring goods from markets to shops by taxi in Kabul is too expensive, and waiting too long in crowded streets makes shop owners out of work.

Agriculture is an important element of the economy. On the one hand, due to rapid urbanization growth and construction of new infrastructure, it has taken a lot of agricultural and valuable land.

4 Feasibility of Smart Transportation

Topic of smart transportation is one of the essential issues for cities that can help to solve most of the problems in the city. Presence of smart control in transportation is to make transportation safe, secure, and environmental-friendly.

In a smart transportation network, sensors and systems are connected to smart devices, which are called internet of things (IoT). As defined by Sethi and Sarangi [19], IoT is a paradigm in which objects equipped with sensors, actuators, and processors communicate with each other to serve a meaningful purpose. In simple words, in an IoT infrastructure, devices share information with users and to a municipality for proper management of transportation.

Smart transportation can reduce unnecessary commuting and extra costs and resources that contribute reduction of greenhouse gas emissions. This type of transport can be cheaply managed by relevant government body. Besides, it can reduce consumption of energy resources by providing information to drivers for choosing the best traffic route with congestion. Some recommended smart transportation features for Kabul are discussed. Street solar lighting with the additional function of power supply for surveillance cameras and Wi-Fi routers, is feasible to invest. Traffic lighting can be managed and controlled smartly by a central command center that keeps control of every traffic light. This system is smart enough to analyze traffic demand and operate in a predetermined manner. This option can be a quick strategy for managing urban traffic and can reduce congestion, prevent accidents, and provide smooth flow for traffic. In each street lighting fixture, free Wi-Fi routers can be installed to help citizens send a notification to near taxi car or bus or for drivers to find the low crowded streets and to find empty car ground to park their vehicles.

On the two sides of the road, smart parking can be designed for medium and short-time parking. Each car ground has a sensor that connects to an app installed on the mobile when this ground is empty. It sends a notification to the mobile app for where the owners can park their cars.

The public transportation in Kabul city can be proposed in two forms: fleet operation concept (working under regular schedule, specific line, and destination such as bus and trolley bus), and individual operation concept (not working under regular schedule, specific line and destination such as minibus, taxi). The fleet operation concept or fleet management can be a smart transportation feature that prepares time schedules and destinations for public transportation. It organizes and regulates the movement of vehicles in specified time and destination.

Mobile applications can be used for guiding tourists, taxi, online shopping, and emergency apps (fire, police, and ambulance). It can contribute to reduce unnecessary commute of citizens and reduce extra costs. Smart cameras can be used for security and law enforcement purposes and traffic management. Smart camera systems can have intelligent control over vehicle behavior and vehicle speed and can track every traffic.

Still, there are challenges for proposing smart technology in transportation and mobility. Lack of knowledge and poor behaviors can be partially solved by providing training and enforcing rules and regulations. This transition needs enough funds, and required technologies and equipment need to be imported from abroad. The introduction of smart technology will need external support in the short run, in the long run, the government should produce revenue to maintain the system.

5 Recommendations

Afghanistan is a developing country that cannot afford high-cost solutions to improve transportation. So it needs low-cost strategies to improve the current situation. Upgrading and improving existing infrastructures and providing effective public transportation. That should be aligned with sustainability requirements within an affordable strategy for Kabul city. Nonmotorized vehicle option is recommended for Kabul city such as bicycle and walking, which are cheap means of transportation.

Considering a lane for bicycles and a walking path are essential. Also, safety of vulnerable users of footpath and streets (children, elder, and disabled people) and modes such as walking and bicycle should be prioritized during planning new mobility and transport mechanisms.

There are lots of vendors on the street, especially in commercial areas. They are vending either on pedestrians or on road shoulders that lead to heavy congestion. So relocating street vendors to a specific location is another solution for encouraging walking and cycling. Consideration must be given to the social aspects of the vendors. They are a good source of income for low vulnerable families.

In the long term, various projects can be proposed to enhance public transport and regulate traffic more effectively to improve the connectivity and accessibility within Kabul city. The proposed projects include widening of community streets, construction of public transport stands and transit stops along roads, and construction of car parks in commercial areas. Another cheap solution for reducing car-dependency is upgrading public transport and providing a bus rapid transit. This option needs informing citizens from public bus destination, schedule, arrival time through information-board or brochures, bus line map, and transportation apps. An emergency line for the fire station and ambulance should be considered as well.

6 Conclusion

This study evaluates the current situation of transportation in Kabul city from social, environmental, and economic aspects. There is inadequate reliable real-time data to analyze the current situation and to propose new infrastructure accordingly. Improvement of transportation sector is directly linked to good health, reduction of poverty, affordable and clean energy accessibility, which are contributing factors to SDGs. A well-designed mobility system with smart technology can improve society's environment, health, economy, and security conditions. Shifting to low carbon mobility is challenging as most of the energy for mobility comes from oil. Two strategies are recommended for the short term: (a) improvement of infrastructure to support low-cost transport means, and (b) fostering an efficient public transportation system. Effectiveness of a transportation system should consider based on accessibility to and affordability by all users, travel time reliability, safety, security, cleanliness, and multimodality. In the long term, use of smart technology can enhance efficiency of mobility and transportation.

References

1. Naganathan, H., Chong, W.K.: Evaluation of state sustainable transportation performances (SSTP) using sustainable indicators. Sustain. Cities Soc **35**, 799–815 (2017). https://doi.org/10.1016/j.scs.2017.06.011
2. Black, W.R.: Sustainable transportation: a US perspective. J. Transp. Geogr. **4**, 151–159 (1996). https://doi.org/10.1016/0966-6923(96)00020-8
3. Organisation for Economic Co-operation and Development (OECD): CO_2 emissions from fuel combustion 2011. https://www.oecd-ilibrary.org/environment/co2-emissions-from-fuel-combustion-2011_co2-table-2011-1-en. Last Accessed 01 Nov 2019
4. Danish, M.S.S., Senjyu, T., Ibrahimi, A.M., Ahmadi, M., Howlader, A.M.: A managed framework for energy-efficient building. J. Build. Eng. **21**, 120–128 (2019). https://doi.org/10.1016/j.jobe.2018.10.013
5. Steg, L., Gifford, R.: Sustainable transportation and quality of life. J. Transp. Geogr. **13**, 59–69 (2005). https://doi.org/10.1016/j.jtrangeo.2004.11.003
6. Shi, Y., Arthanari, T., Liu, X., Yang, B.: Sustainable transportation management: Integrated modeling and support. J. Cleaner Prod. **212**, 1381–1395 (2019). https://doi.org/10.1016/j.jclepro.2018.11.209
7. United Nations (UN): Sustainable Development Goals (SDGs), https://sustainabledevelopment.un.org/sdgs. Last Accessed 01 Nov 2019
8. General Directorate of Traffic—Ministry of Interior Affairs MoI—Afghanistan: Kabul traffic statistics. https://old.moi.gov.af/en/page/directorates/general-directorate-of-traffic (2019)
9. United Nations ESCAP: Economic and Social Survey of Asia and the Pacific 2006: Energizing the global economy. United Nations (UN), New York, USA (2006)

10. Kaijser, A.: How to describe large technical systems and their changes over time? In: Jönson, G., Tengström, E. (eds.) Urban Transport Development: A Complex Issue. pp. 12–19. Springer, Berlin, Heidelberg (2005). https://doi.org/10.1007/3-540-27761-7_3.
11. Pajhwok Afghan News: Traffic Problems, Kabul. https://www.pajhwok.com/en/node/506191, Last Accessed 01 Nov 2019
12. Sasaki: Kabul urban design framework. Ministry of Urban Development and Housing, Kabul, Afghanistan (2017).
13. Special IG for Afghanistan Reconstruction: Street in Kabul. (2017).
14. Results, S.: Web results. Transportation system condition in Kabul, Head of Kabul Traffic Department (2019)
15. Weir, K.: Researchers are identifying startling connections between air pollution and decreased cognition and well-being. https://www.apa.org/monitor/2012/07-08/smog (2012)
16. Zweig, J.S., Ham, J.C., Avol, E.L.: Air Pollution and Academic Performance: Evidence from California Schools. National Institute of Environmental Health Sciences (NIEHS). 1–35 (2009)
17. Every breath we take: The lifelong impact of air pollution. Royal College of Psychiatrists, London, UK (2016)
18. Peterson, B.S., Rauh, V.A., Bansal, R., Hao, X., Toth, Z., Nati, G., Walsh, K., Miller, R.L., Arias, F., Semanek, D., Perera, F.: Effects of prenatal exposure to air pollutants (polycyclic aromatic hydrocarbons) on the development of brain white matter, cognition, and behavior in later childhood. JAMA Psychiatry.**72**, 531–540 (2015). https://doi.org/10.1001/jamapsychiatry.2015.57
19. Sethi, P., Sarangi, S.R.: Internet of things: Architectures, protocols, and applications. J. Electric.. Comput. Eng. **2017**, e9324035 (2017). https://doi.org/10.1155/2017/9324035

From Consumers to Producers: Energy Efficiency as a Tool for Sustainable Development in the Context of Informal Settlements

Zahra Sufizada, Ahmad Ajmal Oryakheill, Mohammad Hafiz Kohnaward, Nabila Fazli, Hasina Zadran, Najib Rahman Sabory, and Mir Sayed Shah Danish

1 Introduction

Clean energy resources and technologies deployed in the context of urban planning and transitioning from informal settlements provide better opportunity to eradicate poverty, promote welfare services, mitigate climate changes, and buoyant socio-economic development [1]. According to Iacoboaea [2], the term slum "describe a heavily populated urban area characterized by substandard housing and squalor". In Afghanistan, settlements described as informal include:

- Unplanned neighborhoods, housing in non-residential zones.
- Illegal settlements and housing on public lands.

Informal urban sprawl around Kabul has grown in recent decades. Population growth, weak public institutions, outdated planning regulations, irresponsive financial systems, and dysfunctional administrative and legal systems are the main factors that have driven the emergence of informal settlements. The Ministry of Urban Development and Land reports 479,972 informal settlements house 80% of Kabul's urban population. Consumption, use, and access to energy are critical issues that impact the quality of life in these settlements, and it impacts climate change mitigation efforts. Also, issues related to environmental pollution and hygiene in rural and informal settlements are utilization of solid waste in form of primary sources of energy. Poorly recycling and consumption primary sources of energy will result in ill-impacts of climate changes and respiratory diseases [3].

Z. Sufizada (✉) · A. A. Oryakheill · M. H. Kohnaward · N. Fazli · H. Zadran · N. R. Sabory · M. S. S. Danish
Kabul University, Kabul 1006, Afghanistan
e-mail: zahrasufizada@gmail.com

M. S. S. Danish
University of the Ryukyus, Okinawa 2030213, Japan

M. S. S. Danish et al. (eds.), *Sustainability Outreach in Developing Countries*,
https://doi.org/10.1007/978-981-15-7179-4_11

Butera et al. [4] described cities as engines for development, cultures, innovations, economies, and relationships. They also identify the worst characteristics of urbanization, such as inequalities and poor living conditions in Latin America and Africa slums. Modern energy had direct benefits on slum dwellers: improved access to lighting, heating, and refrigeration; better access to information; improved indoor air quality; reduction of fire hazards; better education; and improved economic status and investment. Makonese et al. [5] conducted a survey on 52 households on three informal settlements in Tambisa, Johannesburg and Madela Kufa looked at the household needs for cooking, heating, and lighting. The study area was not connected to a municipal electric grid. Factors impact consumption of energy includes seasonal climate, fuel availability and price, and socio-cultural factors that affect the choice and amount of fuels consumed.

Naidoo et al. [6] look at domestic fuel consumption in unelectrified low-income settlements in South Africa. Their findings are based on questionnaires and dispersion modeling for different kinds of emissions. Their study provides a technical dataset that finds that in both electrified and non-electrified areas, solid domestic fuels are still consumed. Another study on energy provision [7] introduces the role of urban slums in the urbanization process using case studies on Kenya, Mali, South Africa, India, and Peru. Electrification, converting waste to energy, and use of new technologies are suggested as solutions. Recent evidence focused on the concept and practical solutions related to energy to waste based on 4 E (energy, exergy, environment, and economy [8]. That followed by a number of studies involving leveling of energy provision cost [9], mitigate climate changes and limit the temperature increase to 1.5 °C preindustrial levels, models (hybrid waste to energy [10], biogenic carbon neutrality model [11]), approaches and frameworks [12, 13].

An estimated one billion people live in slums around the world [4]. Among many global challenges that these slums face, access to energy is one of the most important. As the urban population grows, infrastructure resources face increasing pressure, and problems with access to energy for slum residents are complex and context-driven. A major source of urban pollution comes from the domestic combustion of fuels in informal settlements. Low quality and nonrenewable energy sources such as wood, coal, and kerosene are used to meet the energy needs of poor informal residents [5]. Incinerating these materials in poor conditions causes or worsens respiratory, pulmonary, and systematic diseases [14]. As a solution, improving access to clean and modern energy sources such as electricity, clean fuels, and clean cooking technologies is important for improving health and education [15].

This research investigates the energy consumption and use in informal settlements to develop efficient solutions in terms of energy use, combustion material, and building materials. These proposals should address inequalities among citizens, access to clean water and sanitation, access to clean and modern energy, improved health of citizens, and development of sustainable communities. This study aims to clarify the importance of energy consumption and energy efficiency in informal settlements. This paper also proposes solutions for type of combustion materials, infrastructure improvement, and building conditions.

2 Methodology

This analysis is based on a literature review, case studies of specific areas, and development of the questionnaire. That follows the rationale of other studies and focuses on the topics of energy access, energy use, and energy efficiency of informal settlements. The proposed framework is similar to many previous studies about communities in North America, Kenya, Mali, India, Peru, and South Africa, which is adapted for Kabul. The Questionnaire Survey was designed in November 2019 to gather information in informal settlements on selected issues such as:

- Access to energy sources (electrification, heating, cooling, lighting).
- Quality of buildings in terms of orientation, open spaces, and materials.
- Quality of living in informal neighborhoods in terms of access to primary infrastructures (Fig. 1).
- Quality of local transportation and settlement (Fig. 2).

The survey contains 22 closed and open-ended questions. This questionnaire does not include personal information from the respondents. Our team selected the houses randomly. It took 2 days to administer 25 questionnaires.

Fig. 1 Qala-e-Zaman Khan informal settlements (December 12, 2019)

Fig. 2 Qala-e-Zaman Khan local alleyway view (Dec 12, 2019)

2.1 Case Study

The survey was conducted in one informal settlement area: Qala-e-Zaman Khan, 16th District, Kabul (Fig. 3).

Fig. 3 Cast study location (Qala-e-Zaman Khan) [16]

2.2 Data Analysis

Our methodology is informed from earlier case studies that guide us in developing solutions. In studying Qala-e-Zaman Khan, our team visited the site to document its existing condition. This site visit helped us identify the main challenges to address in the questionnaire (Figs. 4 and 5, Table 1). This informal settlement has access to the municipal electric grid, and the primary use is for lighting. Little electricity was used for heating and cooling as residents did not have modern technologies such as air conditioning. Homes without electricity used lanterns for lighting while wood and coal for used for heating and cooking (Fig. 6). The water quality of the area was good as each household has access to its own well for drinking water. Few were able

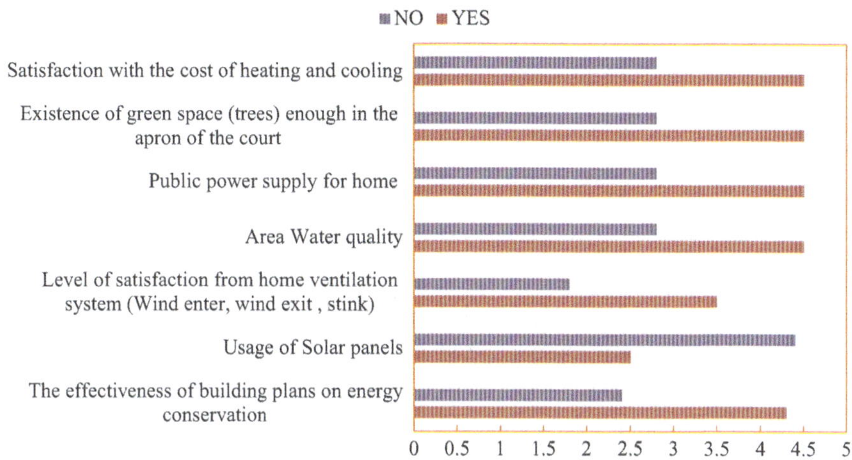

Fig. 4 Results of the first sort of questions (yes–no)

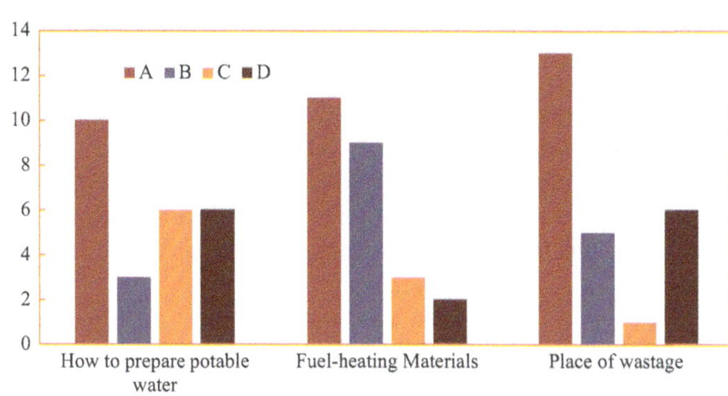

Fig. 5 Results of the second sort of questions (optional questions)

Table 1 The 4 option questions details

No	Option A	Option B	Option C	Option D
1	Having well	Contract with private company	Buying water from tankers	Having general wells in place
2	Wood	Coal	Plastic	Leather
3	Private garbage	by vehicles	outside garbage	Landfills

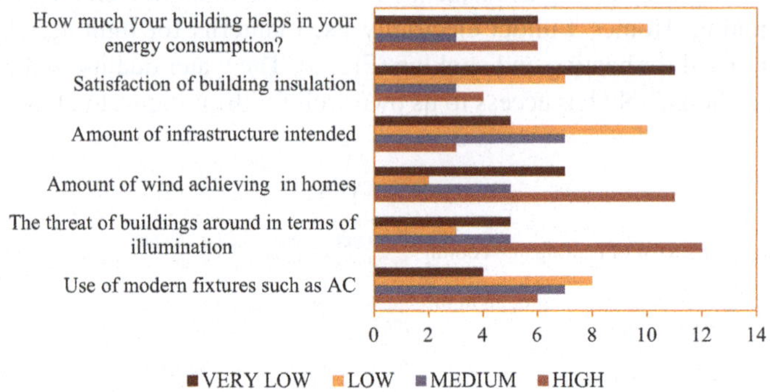

Fig. 6 Results of the second sort of questions (energy consumption and efficiency-related questions)

to access solar energy. The construction and orientation design of buildings were poor, and often one house blocks the sun and view of a neighboring house.

Here are the results of the interview questions:

- The costs for heating and cooling of non-planned homes are very high compared to the planned homes because they are not designed according to the building standards.
- Most of the building's walls and roof coatings are made of traditional materials that do not promote energy conservation.
- Nonrenewable energy sources are used to cool and heat buildings at a high cost and contribute to environmental damage.
- Many features differ between the planned homes and non-planned homes, including the non-standard design of buildings, high energy consumption, inappropriate orientation, and lack of appropriate infrastructure.
- Usually wood is used for the construction of windows and gates; compared with modern materials, resulted in higher heating and cooling costs.

3 Recommendations

Building on our analysis of previous research and the questionnaire results, we developed some solutions that are divided into three main categories that also address the SDGs:

- Building condition (orientation, open space, and envelope materials)
- Energy use (for combustion, lighting and other activities)
- Infrastructure.

3.1 Building Conditions

Orientation and open space: Orientation is the positioning of a building in relation to seasonal variations in the sun's path and prevailing wind patterns. Proper orientation can increase the energy efficiency of houses, making it more comfortable and cheaper to live [17]. Carefully orienting the building to the south, maximizing the south face view, minimizing east and west openings, and using plants are common strategies that can make building orientation more comfortable in different climatic conditions. Maximizing the use of passive energy strategies can help to develop efficient housing and neighborhoods [18]. However, re-orienting informally-built structures are extremely challenging, and those buildings may need to be demolished. These strategies can be applied when redesigning informal communities into more sustainable neighborhoods.

Many of Kabul's informal settlements have problems with unlighted streets, no spaces for social activities, and a lack of green space. Open space can shape an urban neighborhood's social and cultural identity and can improve the local environment by reducing air pollution. Open space in many informal neighborhoods are unused or are dumpsites. We can use these spaces to upgrade the living quality of informal neighborhoods and provide open green spaces for the residents, as shown in Figs. 7 and 8 [18].

Fig. 7 Design for climate and winter sun

Fig. 8 Design for summer shed and ventilation

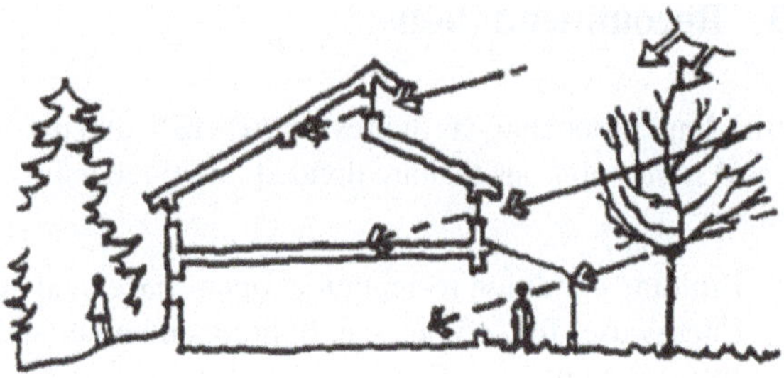

3.2 Building Envelope

To maximize a building's energy efficiency, ecological materials should be used [19]. Sheep wool is one of the best options for insulation and has the following desired characteristics:

- Sheep wool does not cause irritation to the eyes, skin or lungs.
- Sheep wools are breathable (it can absorb and desorb moisture).
- Wool is not combustible.
- Sheep wool absorbs and reduces noise.

Korjenic et al. [19] showed that thermal insulation made from sheep wool had comparable or superior performance to insulation made from conventional materials (Fig. 9). Sheep wool can be recommended for renovating existing informal settlements and is abundant in a developing country like Afghanistan. As a result, increased use of sheep wool insulation also helps improve indoor air quality and improve the country's economy.

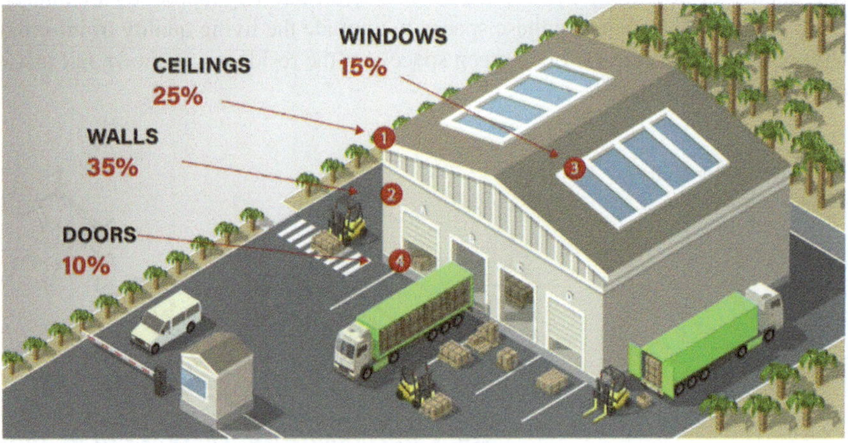

Fig. 9 A surmise of air loss due to poor insulations [20]

- Improving the energy efficiency of building roofs has a positive effect on the energy efficiency in buildings:
- Adequate attic insulation can be added to the existing roof by considering different properties of roof insulation.
- Cooling roof coatings can reduce heat retention, can increase the U value, and extends the life of the roof. Polyurethane foam works best on flat or low sloped roofs.
- Energy-efficient roofing materials. the most popular options include:
- Asphalt shingles, shake shingles, metal roofing, slate roofing, tile roofing.

Harvesting and storing rainwater also contributes to energy-efficiency and sustainability. Important factors include selecting appropriate roof materials, avoiding the usage of biocides, algaecides, and fungicides, and clearing the gutters and roofs of leaves.

3.3 Combustion Material

Domestic fuel use in low-income informal settlements consumes vast amounts of wood, coal, kerosene, animal dung, electricity, and urban wastes for energy needs and is a major contributor of urban air pollution in Kabul. Liquid gas is used for cooking, while solid fuels such as wood and coal are preferred for heating in the winter. Paraffin lanterns are used for lighting purposes [4]. Factors such as seasonality, availability, price of fuels, cultural preferences can affect residents' fuel choices and consumption quantity [5].

According to our analysis, Qala-e-Zaman Khan's residents preferred wood and coal as primary combustion fuel. Other poorer Kabul slums that do not have access to electricity choose wastes as combustion material. Based on our analysis and similar surveys from different countries, authors following suggestions considering in terms of the following solutions.

3.4 Electrification

Electrification of informal settlements plays a vital role in improving residential efficient energy usage. For lighting, energy is cheaper and cleaner than biomass or kerosene fuels. Wastes could be converted into electrical energy if health, environment, and livelihood impacts are considered. Butera et al. [4] list two ways that wastes should ideally be converted to energy:

- The combustibles can be removed, concentrated or condensed, and sold as a fuel to supplement or replace wood.
- Organic components can be used in digesters to produce natural gas.

Rural areas in developing countries are strongly encouraged to use new technologies and renewable sources of energy like biomass and solar energy.

3.5 Infrastructure

Informal settlements grew after the war as citizens moved to urban areas like Kabul to seek new opportunities. These slums grew haphazardly with a lack of proper infrastructure, unsteady and unsafe housing, and poor sanitation. Flooding and accumulation of stagnant water are major problems in Kabul's informal settlements due to its low-lying topography, the closeness of shacks along dirt roads, and a lack of stormwater drainage systems. Excessive flooding, especially in informal settlements in Qala-e-zaman Khan, displaced over thousand people each year. Undrained water seeps into houses and causes significant property and personal damage, such as the degradation of floors, walls, and personal belongings, including mattresses and clothes. Standing water mixed with high levels of ground contaminants becomes a health hazard.

A sustainable urban drainage system (SUDC) has been proposed to reduce the flow of stormwater from unwanted areas. A globally-accepted solution, the main interventions of SUDC involve planning and building swales, soak ways, infiltration trenches, and wetlands. These solutions can be empowered by including community members to implement and manage stormwater solutions [21]. Swales are narrow culverts dug into the ground in low-lying areas with shallow dish-shaped depressions. These swales are covered by vegetation to redirect the flow of water away from unwanted areas. Soak ways and infiltration trenches redirect the water runoff through a filtration process with layered rocks. Soak ways have biofilters that absorb nutrients. In both solutions, the filtered water is redirected to a wetland, which is rich with biological variety. Creating a barrier to redirect the water path helps in the absorption of water to the ground. All these systems can be effective solutions relative to their size and location [21].

4 Relevance to Kabul Urban Design Framework (KUDF)

Organic Neighborhoods are mentioned in the Kabul Urban Design Framework as one of the most important projects that should be developed [22]. There are different strategies in KUDF for the improvement of these settlements in terms of infrastructure, land, open spaces, and relocation. The design strategies mentioned in KUDF are.

Fig. 10 Site area before public realm upgrading [22]

4.1 *Public Realm Upgrading*

In neighborhoods with poor access to transportation, infrastructure, or urban amenities (Fig. 10) there should be a focus on paving roads and improving infrastructure (Fig. 11).

4.2 *Community Land Trust*

Community land trusts should facilitate community-level planning and ownership of infrastructure, public amenities, and housing (Figs. 12 and 13).

Identify key sites and corridors to help neighborhoods face increasing development pressure through a sensible redevelopment strategy (Figs. 14 and 15).

The expansion of the Kabul's electric grid is one of the important goals identified by Sasaki. KUDF determined that population growth, higher demand, and the increased use of electric appliances will result in a large increase in demand. The increased reliance on the electric grid offers an opportunity for Kabul to mitigate

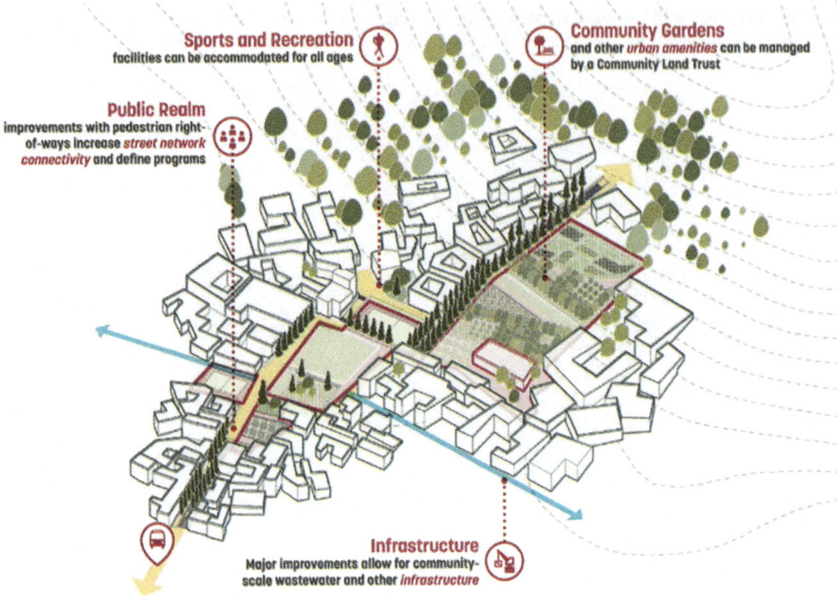

Fig. 11 Site area after public realm upgrading [22]

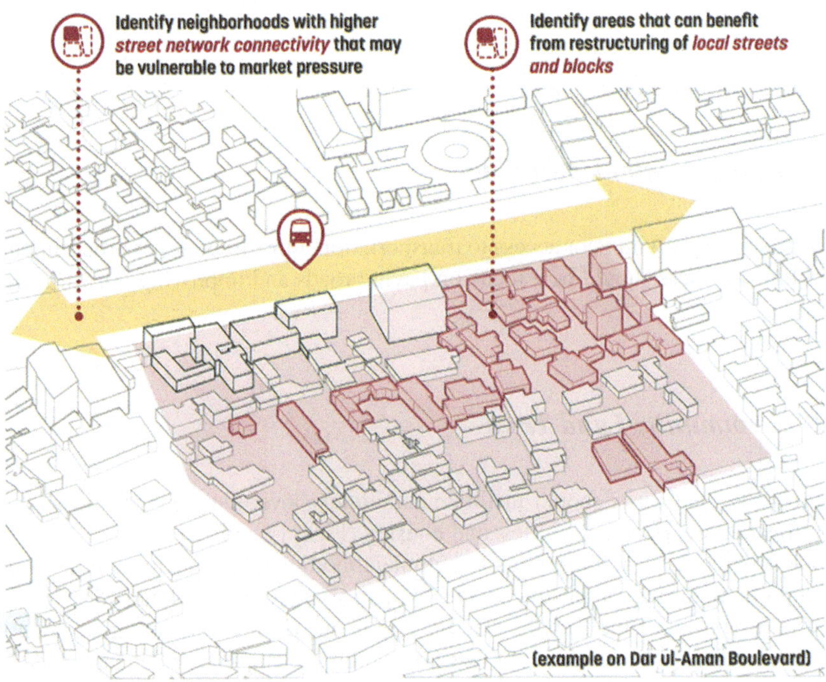

Fig. 12 Site area before community land trust [22]

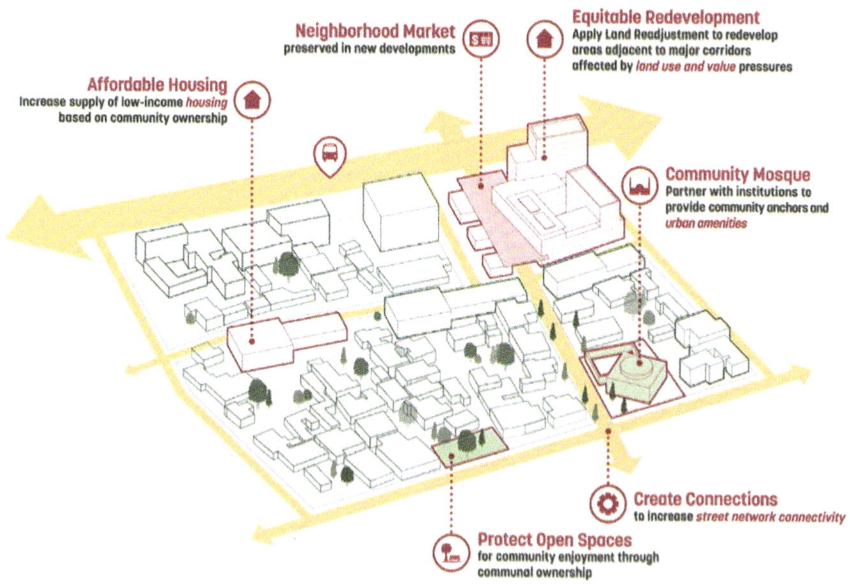

Fig. 13 Site area after community land trust [22]

carbon emissions as the economy grows. Sasaki also considers energy security as the basis of sustainable energy plans and improvement of energy efficiency. To help Kabul achieve its goals to become a green and sustainable city, Sasaki presents these concepts as solutions: expanding the electric grid, expand regional renewable energy, develop solar energy, and promote energy efficiency technologies. The concepts proposal is based on these general principles: promoting locally distributed renewables, improve energy efficiency, promote low carbon sustainable development, and universal electricity access.

To expand the electric grid, Sasaki suggests building new substations on vacant land and increasing the capacity of existing substations around Kabul for regional energy improvement. Wind energy turbines can be installed in hilltop and agricultural areas to maximize energy generation using seasonal winds. The use of rooftop PV solar panels can independently power different kinds of buildings and relieve energy demand. Further proposed energy efficiency improvements throughout Kabul include building efficiencies, LED lighting programs, circular energy and economy programs, and a smart grid system.

We conclude that KUDF provides a general framework to improve urban slums using general principles and design concepts that can be applicable to Kabul but can be modified according to context. Our solutions are generally related to Sasaki but with some differences in building quality solutions.

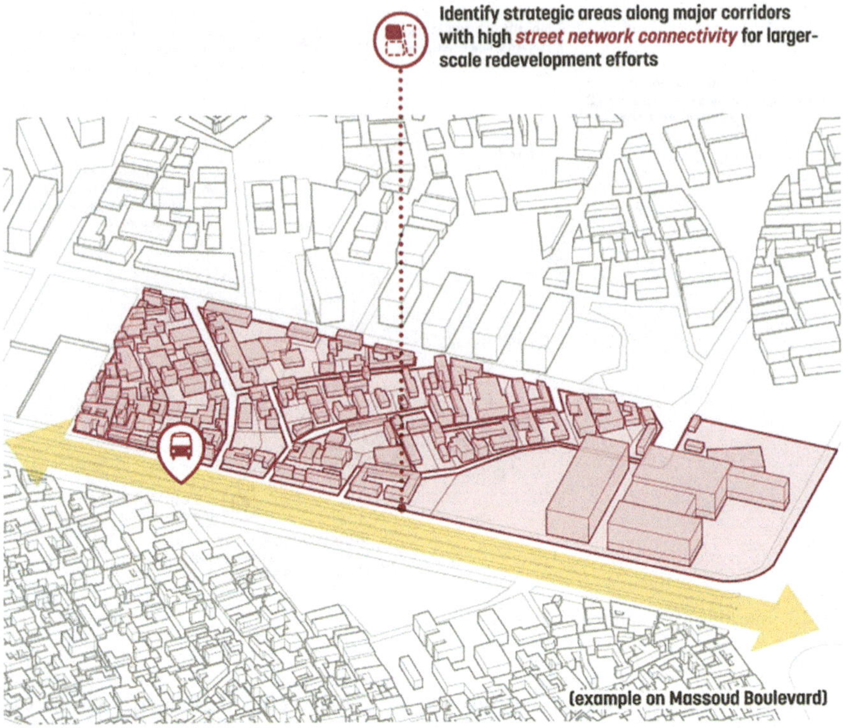

Fig. 14 Site area before redevelopment [22]

5 Relevance to Sustainable Development Goals (SDGs)

The purpose of the paper is to define strategies and solutions to help Kabul meet sustainable development goals (SDGs). Our proposed solutions address six goals:

5.1 First Goal: End Poverty in All Its Forms

Around the globe, 800 million people live in extreme poverty, those live in urban informal settlements, mostly do not have access to many urban services [23]. By supporting them with access to clean water, stormwater management, new and clean energy technology, and upgraded spaces, the government can end poverty in these settlements. Poverty reduction and economic buoyant require exhaustive policies within workable strategies for a long run sustainability [24]. Researches highlight the emerging role of poverty reduction and socio-economic sustainability in the context of economic frameworks and development strategies that reinforce access to energy-efficient services and welfare [25].

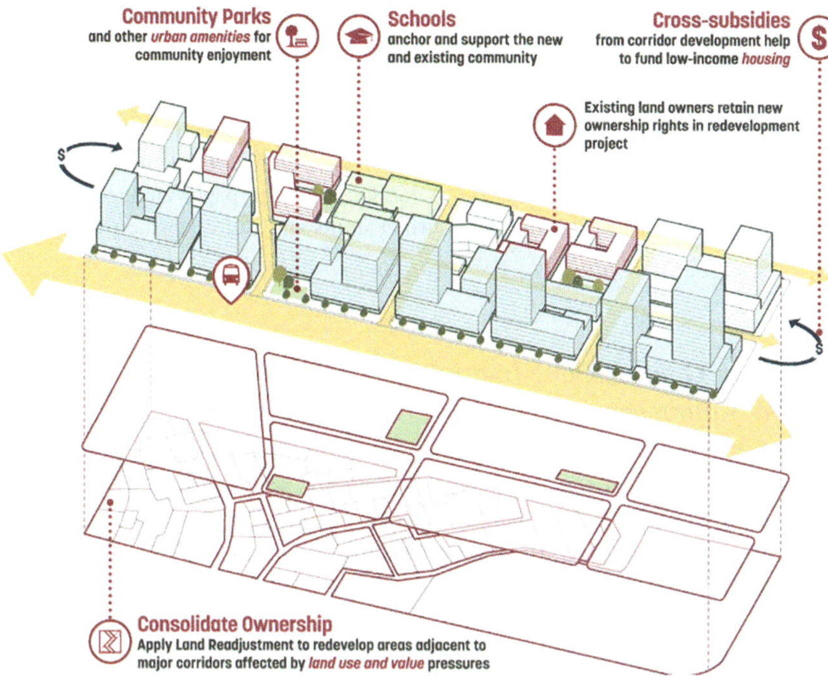

Fig. 15 Site area after redevelopment [22]

5.2 *Third Goal: Ensure Healthy Lives and Promote Well-Being for All Ages*

Reducing air pollution alleviates many health problems common among the poor residents of informal settlements. Our study connects the pollution caused by the inefficient use of energy with poor living conditions of urban slums, which lead to health problems. Using clean energy resources and developing green, open spaces should improve the health and well-being of all residents. There are growing approaches in the context of sustainable development of developing countries from energy accessibility point of view.

5.3 *Sixth Goal: Ensure Availability and Sustainable Management of Water and Sanitation for All*

Clean and accessible water is essential for all humankind, but poverty and insufficient water infrastructure impact health and life expectancy due to diseases caused by water scarcity, poor water quality, and inadequate sanitation Improving water quality in informal settlements requires providing access to clean water, treating polluted water, and managing stormwater.

5.4 Seventh Goal: Ensure Access to Affordable, Reliable, Sustainable, and Modern Energy for All

From 1990 to 2010, number of people with access to electricity increased to 1.7 billion, and investment in clean energy resources such as wind and solar energy can further increase affordable access to energy [23]. Our paper analyzes the type of fuels used for household activities in urban slums and addresses their use of new technologies.

5.5 Tenth Goal: Reduce Inequality Within and Among the Countries

While citizens living in poverty have no difficulty accessing essential services, inequities often differ between different countries as well as within a country's borders [23]. Our solutions try to raise the standard of living for slum residents to one comparable to the average citizen.

5.6 Eleventh Goal: Make Cities and Human Settlements Inclusive, Safe, Resilient, and Sustainable

As hubs for ideas, cultures, productivity, and social development, cities face challenges with providing housing and having adequate budgets to maintain municipal services and infrastructure while planning for its future [23]. Kabul should ensure its services can be accessible to its slums to improve the city as a sustainable and safe environment.

6 Suggestions

The following solutions address their relevance to Leadership in Energy and Environmental Design (LEED) certification. The Sustainable Sites category is focused on balancing the proper density of buildings with surrounding open, green spaces. As mentioned in Sasaki, one can vertically develop buildings and decrease disturbing open spaces. This will allow the planning of appropriately accessible roads throughout the settlement.

Energy efficiency: our solutions recommend properly orienting and arranging plots by using natural shading devices as trees. Using sheep wool as insulation should increase the energy efficiency of newly constructed buildings.

Renewable energy generation: our proposal to rely on renewable resources such as wind and solar can improve the energy efficiency and usage of buildings in the neighborhood. Materials and resources: Our proposal to using local sheep wool for insulation demonstrates the use of a local, renewable, and certified product as an environmentally friendly material that can contribute to domestic economic growth.

7 General Recommendations

- Reduce access to charcoal and wood to encourage using clean energy sources for slum residents.
- Fuel subsidy and tax policies shall be consistent with policy goals and ensure affordable, equitable access. s.
- Encouraging people to save their environment can be a significant step towards achieving SDGs goals.
- Provide financing and credits for rewiring electrical connections inside homes.
- Decentralize renewable energy sources and technologies to widen access to citizens.
- Finance, subsidize, or incentivize purchasing appliances that use renewable energy. Low-cost water heaters are mentioned in KUDF.
- Develop policies promoting energy efficiency, especially using locally sourced sustainable resources.
- Develop green, open spaces to organize layout in slum areas better and enhance the cultural, educational, and social environment.
- Improve water quality and access in slums through water management policies and infrastructure.
- Promoting, educating, and encouraging citizens about sustainable buildings, communities, and energy.

8 Conclusion

As one of the main consumers of energy in Kabul, informal settlements/slums have significant challenges with urban development. Lack of access to public facilities and services, infrastructure, clean water, and clean energy, these settlements have poor building orientation, and construction impedes progress to improve the quality of living and to meet the Sustainable Development Goals for future urban development. We present general strategies for improving these settlements, followed by specific solutions appropriate for urban slums. Ultimately the improving energy use, infrastructure, and building conditions can begin a transition within informal settlements from consumers to producers. This paper provides initial steps in raising interest in studying energy use patterns in Kabul's informal settlements. Relying on research from different countries, we developed some solutions for upgrading urban slums.

These solutions are focused on the urban design of the neighborhoods. Although Sasaki provides a general framework and policy recommendations for energy and urban upgrading, our research placed significant focus on energy use for urban slums with a broad vision that included the importance of building quality and quality of life. It is hoped this research approach can be adopted in future investigations.

References

1. Danish, M.S.S., Sabory, N.R., Ershad, A.M., Danish, S.M.S., Yona, A., Senjyu, T.: Sustainable architecture and urban planning trough exploitation of renewable energy. Int. J. Sustain. Green Energy. **6**, 1–7 (2017). https://doi.org/10.11648/j.ijrse.s.2017060301.11.
2. Iacoboaea, C.: Slums in Romania. Theor. Empirical. Res. Urban Manag. **4**, 101–113 (2009)
3. Danish, M.S.S., Senjyu, T., Ibrahimi, A.M., Ahmadi, M., Howlader, A.M.: A managed framework for energy-efficient building. J. Build. Eng. **21**, 120–128 (2019). https://doi.org/10.1016/j.jobe.2018.10.013
4. Butera, F.M., Caputo, P., Adhikari, R.S., Facchini, A.: Urban development and energy access in informal settlements. A review for Latin America and Africa. Procedia Eng .**161**, 2093–2099 (2016). https://doi.org/10.1016/j.proeng.2016.08.680.
5. Makonese, T., Masekameni, D.M., Annegarn, H.J.: Energy use scenarios in an informal urban settlement in Johannesburg, South Africa. In: 2016 International Conference on the Domestic Use of Energy (DUE). pp. 1–6 (2016). https://doi.org/10.1109/DUE.2016.7466703.
6. Naidoo, S.: Domestic fuel combustion in unelectrified low-income settlements in South Africa. In: 2015 International Emission Inventory Conference—"Air Quality Challenges: Tackling the Changing Face of Emissions. pp. 1–21. United States Environmental Protection Agency (EPA), San Diego (2015)
7. Droege, P.: Urban Energy Transition: From Fossil Fuels to Renewable Power. Elsevier, Amsterdam (2011)
8. Carneiro, M.L.N.M., Gomes, M.S.P.: Energy, exergy, environmental and economic analysis of hybrid waste-to-energy plants. Energy Convers. Manage. **179**, 397–417 (2019). https://doi.org/10.1016/j.enconman.2018.10.007
9. Vaillancourt, K., Bahn, O., Levasseur, A.: The role of bioenergy in low-carbon energy transition scenarios: A case study for Quebec (Canada).Renew. Sustain. Energy Rev. **102**, 24–34 (2019). https://doi.org/10.1016/j.rser.2018.11.025
10. Zhang, J., Kan, X., Shen, Y., Loh, K.C., Wang, C.H., Dai, Y., Tong, Y.W.: A hybrid biological and thermal waste-to-energy system with heat energy recovery and utilization for solid organic waste treatment. Energy **152**, 214–222 (2018). https://doi.org/10.1016/j.energy.2018.03.143
11. Hugues, P., Assoumou, E., Maizi, N.: Assessing GHG mitigation and associated cost of French biofuel sector: Insights from a TIMES model. Energy **113**, 288–300 (2016). https://doi.org/10.1016/j.energy.2016.06.146
12. Katinas, V., Marčiukaitis, M., Perednis, E., Dzenajavičienė, E.F.: Analysis of biodegradable waste use for energy generation in Lithuania. Renew. Sustain. Energy Rev. **101**, 559–567 (2019). https://doi.org/10.1016/j.rser.2018.11.022
13. Danish, M.S.S., Senjyu, T., Zaheb, H., Sabory, N.R., Ibrahimi, A.M., Matayoshi, H.: A novel transdisciplinary paradigm for municipal solid waste to energy. J. Cleaner Prod. **233**, 880–892 (2019)
14. Gordon, S.B., Bruce, N.G., Grigg, J., Hibberd, P.L., Kurmi, O.P., Lam, K.H., Mortimer, K., Asante, K.P., Balakrishnan, K., Balmes, J., Bar-Zeev, N., Bates, M.N., Breysse, P.N., Buist, S., Chen, Z., Havens, D., Jack, D., Jindal, S., Kan, H., Mehta, S., Moschovis, P., Naeher, L., Patel, A., Perez-Padilla, R., Pope, D., Rylance, J., Semple, S., Martin, W.J.: Respiratory risks from household air pollution in low and middle income countries. Lancet. Respir. Med. **2**, 823–860 (2014). https://doi.org/10.1016/S2213-2600(14)70168-7

15. Heltberg, R.: Factors Determining Household Fuel Choice in Guatemala. Social Science Research Network, Rochester, NY (2003)
16. Map showing the location of Qala-e-Zaman Khan, https://earth.google.com/web/@34.5255344,69.21857414,1810.9713203a,2092.15660064d,35y,0.00477978h,0t,0r/, (2020)
17. Holmes, G.E., Patterson, J.R., Stalling, J.E.: Sense of place: issues in counseling and development. J. Humanistic Couns, Educ. Dev. **42**, 238–251 (2003). https://doi.org/10.1002/j.2164-490X.2003.tb00009.x
18. Kazimee, B.A.: Sustainable urban design paradigm: twenty five simple things to do to make an urban neighborhood sustainable. WIT Press. 11 (2002)
19. Korjenic, A., Klarić, S., Hadžić, A., Korjenic, S.: Sheep wool as a construction material for energy efficiency improvement. Energies **8**, 5765–5781 (2015). https://doi.org/10.3390/en8065765
20. A guide to energy loss within your home, https://www.climashieldsprayfoam.com/your-energy-bill/. Last Accessed 17 Mar 2020
21. Muniz, E., Jeyaraj, E., Button, K., Ma, R.: Adapting Sustainable Urban Drainage Systems to Stormwater Management in an Informal Setting. Interactive Qualifying Projects (All Years). (2010)
22. Kabul Urban Design Framework: https://www.sasaki.com/projects/kabul-urban-design-framework/, (2018)
23. Sustainable Development Goals (SDGs) booklet.: United Nations Development Programme (UNDP), New York (2017)
24. Danish, M.S.S., Sabory, N.R., Danish, S.M.S., Ludin, G.A., Yona, A., Senjyu, T.: An open-door immature policy for rural electrification: a case study of Afghanistan. Int. J. Sustain. Green Energy **6**, 8–13 (2016). https://doi.org/10.11648/j.ijrse.s.2017060301.12
25. Danish, M.S.S., Sabory, N.R., Danish, S.M.S., Senjyu, T., Ludin, G.A., Noorzad, A.S., Yona, A.: Electricity sector development trends in an after-war country: Afghanistan aspiration for an independent energy country. J. Energy Power Eng. **11**, 553–557 (2017). https://doi.org/10.17265/1934-8975/2017.08.007

Efficient Use of Energy and Its Impacts on Residential Sector: A Step Towards Sustainable Building

Hamid Maliki, Mikaeel Ahmadi, and Najib Rahman Sabory

1 Introduction

An energy-efficient building has many benefits; firstly, it reduces negative impact on the environment and advances in energy security. The adverse effects on the environment include both resource consumption and pollution in building sector. Continuous environmental deterioration and scarcity of available natural resources have raised global concerns on the exploitation of renewable energy sources and the effective application of energy conservation strategies in all energy-consuming sectors [1]. Energy requirement for buildings consumes around 40% of the world's total energy consumption. One-third of the harmful gas emissions around the world is generated by energy used in buildings. Energy demand is increasing due to population growth. Heat gains and heat losses through the building envelope are responsible for high-energy consumption. Improving energy efficiency of the building reduces both gains and losses through the envelope [2]. The high prices of fossil fuel and inefficient use have prompted governments and policymakers to develop strategies to reduce energy consumption and dependence on fossil fuels. However, energy consumption in residential sector is predicted to increase at a rate of 1.1% per year from 2008 to 2035 [3]. Most of the buildings in Afghanistan are very energy inefficient and may struggle to provide basic needs. They are cold in the winter and often experience indoor air pollution from fuel use.

Residential and commercial buildings in Afghanistan consumed about 74% of electrical energy supplied by DABS in 2013–2014 for meeting space heating, cooling, lighting, cooking, water heating, refrigeration, electronics and computer, and other

H. Maliki (✉) · N. R. Sabory
Kabul University, Kabul 1006, Afghanistan
e-mail: malikihamid1996@gmail.com

M. Ahmadi
University of the Ryukyus, Okinawa 903-0213, Japan

M. S. S. Danish et al. (eds.), *Sustainability Outreach in Developing Countries*,
https://doi.org/10.1007/978-981-15-7179-4_12

needs [4, 5]. Afghanistan imports 77% of its energy from the neighboring countries, and residential buildings consume 71% of total energy [6]. Most of this energy is consumed for heating in buildings because we have significant heat loss in winter. Insufficient knowledge of thermal performance of building structure is the main reason behind this situation. Energy EE in buildings goes through the Envelop (insulation, double glazing, passive heating), Geometry system, and behavior of users.

Lack of energy in Afghanistan made people face different challenges in the economic, social, political, and security sides. On the other side, inefficient use of energy in residential sector causes environmental issues and decreases access to the affordable, reliable, sustainable, and modern use of energy supply. High price of electricity 14 AFN/kWh in governmental and commercial buildings is another problem. Some problems which 11th district residents are facing are as follows. First, there is not enough electrical energy. Second, more than half of Afghanistan's energy is supplied by neighboring countries, and lastly, high price of fossil fuels and their impact on environment and ecosystem.

2 Energy Consumption

Role of lighting is vital in daily activities, especially at night and where natural light is not available. As our economic growth and populations expanded, global demand for lighting is also increased. According to the International Energy Agency (IEA), electricity consumed for lighting is about 20% of the world's output power stations [7].

Building energy consumption is among the highest of total energy consumption. So lighting systems are an important aspect of energy saving. Energy-efficient equipment and effective control of the lighting system are the most important ways to prevent wasting electrical energy. Using less electric lighting also reduces the use of air-conditioning energy and improves thermal comfort because using more electric lighting systems is directly proportional to heat gain.

Efficient use of energy in lighting system helps to access affordable, reliable, sustainable, and modern energy for all. Besides these benefits, EE policies are very environmentally friendly. Because there is almost 58% carbon dioxide emissions around the world each year from residential lighting system [8], the increase of carbon dioxide gas (CO_2) and other greenhouse gases make the environment unstable, therefore causing global warming. This not only increases health risks but also increases destruction of the ecosystem [8]. Some trends reduce electric light consumption by applying new technologies. The methods to reduce lighting electricity consumption are use of light sources with high luminous efficacy, minimum possible power density, use of lighting control systems, and utilization of daylight.

Role of other light appliances is important in energy consumption sector, and implementation of EE on light appliances also helps to save energy and contribute to reach the SDG targets. Cooling system in this study is a part of a light appliance (low

Fig. 1 Cooling appliance usage in Kabul city [9]

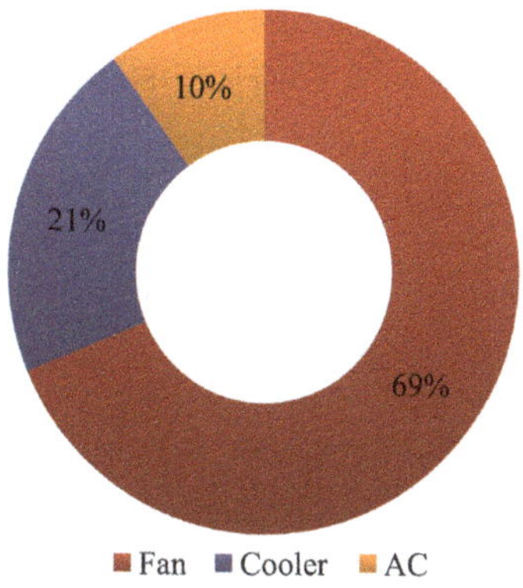

energy appliances) system because in Kabul city, more than 68.79% of buildings are using fans for cooling (Fig. 1).

3 Solar Lighting System

Solar energy is energy from the sun in the form of radiated heat and light. It drives the climate and weather and supports life on earth. Solar energy technologies make controlled use of this energy resource. Solar power generation is a sustainable, clean, and renewable alternative which, if adopted, can protect the planet and the coming generations from harmful environment degradations [10].

In this study, an experimental prototype is recognized, which generates electricity for lighting systems and small appliances such as TV, computers, fans, and smartphones.

3.1 System Design Methodology

The system design firstly requires the installation of a solar panel on the exterior of the building. The output of the solar panel is then connected to a built inverter (containing inverter and charge controller together), which regulated the voltage/current coming from the solar panel and going into the battery. The load is connected to the inverter; in necessity, the battery connects to the inverter.

- Experimental prototype: A small-scale prototype of the system was sized and tested to see its performance. The overall system is prepared using the following main components: Solar PV Panel, Solar Charge controller, DC Battery, Inverter.
- Load calculation: Authors assume an average electric consumer and, after calculation, apply this sample in all of the buildings in the 11th district. The economic

Table 1 Load calculation for an average home

Appliance	Number	Power (W)/unit	Power (W)	Working Hour (h)/day	Energy (kWh)/day
Light LED	10	9	90	6	540
TV	1	80	80	6	480
Laptop	2	40	80	5	400
Smart Phone	3	5	15	4	60
Fan	3	35	105	5	525
Total power demand			370	Total demand/day	2005

analysis and payback period are also recognized by the base of this sample shown in Table 1.

PV panel sizing: The inverter efficiency is 93% and cable efficiency should be 98%. Therefore, the total power is as follows:

$$\begin{aligned}\text{Total Power} &= \frac{\text{proposed power}}{(\text{inverter efficiency})(\text{cable efficiency})}\\ &= \frac{370\,\text{W}}{0.93 \times 0.98} = 406\,\text{W}\end{aligned} \tag{1}$$

So, number of modules is calculated as follows:

$$\text{Number of modules} = \frac{\text{total power}}{\text{module power}} = \frac{406\,\text{W}}{160\,\text{W}} \approx 3 \tag{2}$$

The main task of an inverter is to convert direct current (DC) to alternating current (AC). The inverter should also provide safe operation and high efficiency. The efficiency of an inverter varies with input power and voltage. The inverter is sized by maximum power and stirring voltage. Power is calculated as follows:

$$\begin{aligned}\text{P}_{\text{inverter}} &= \frac{(\text{Proposed demand power})(\text{safety factor})}{\text{inverter efficiency}}\\ &= \frac{(370\,\text{W})(1.2)}{0.93} = 478\,\text{W}\end{aligned} \tag{3}$$

The number of inverters is calculated as follows:

$$\begin{aligned}\text{Number of inverters} &= \frac{\text{P}_{\text{inverter}}}{\text{Inverter selected power}}\\ &= \frac{478\,\text{W}}{720\,\text{W}} = 0.6 \approx 1\ \text{unit}\end{aligned} \tag{4}$$

Table 2 Initial cost analysis of PV system

Component	Quantity	St/unit (AFN)	Total cost (AFN)
Modules	3	4,000	12,000
Inverter with MPPT	1	5,500	5500
Battery	2	3,000	6000
Total initial cost			18,500
Total cost			703,000,000 ≅. 9,165,000

Battery backup sizing: Battery sizing consists of determining the capacity of the battery (Ampere-hour), the voltage of the battery (Volt), and the type of the battery (ordinary battery or deep cycle battery). Sizing of the battery is such that the battery should be large enough to store sufficient energy to operate the appliances for a specific time. The amount of energy that should be produced by a battery, as found before, is 76.053 kWh per day.

$$\begin{aligned}\text{System cpacity (Ah)} &= \frac{\text{Demanded energy}}{\text{efficiency} \times \text{DOD} \times \text{Batt.Syst.Volt}} \\ &= \frac{2005\ \text{Wh}}{0.9 \times 0.7 \times 12\ \text{V}} = 265.2\ \text{Ah}\end{aligned} \tag{5}$$

Now we obtain the number of batteries:

$$\begin{aligned}\text{Number of batteries} &= \frac{\text{System cpacity (Ah)}}{\text{Battery Capacity}} \\ &= \frac{265.2\ \text{Ah}}{150\ \text{Ah}} = 1.7 \approx 2\end{aligned} \tag{6}$$

Cost analysis: In this section, we will discuss initial cost, annual energy saving, and the payback period of this project and review if applying this project is effective. The implementation cost of the PV system according to Afghan company prices is shown in Table 2.

As we mentioned before, this system is a sample for an average energy consumer building in the 11th district and if we find the total cost of the project, the initial cost of building multiply by the total number of buildings.

3.2 Annual Saving

To find the annual benefit of PV system, we first calculate the amount of energy that the system can generate annually [11]. Based on having 300 sunny days in Kabul city now we have the following:

$$\begin{aligned}\text{Energy generation per year} &= (\text{PVs capacity power})\left(\frac{\text{generation days}}{\text{year}}\right)\\ &= (2\,\text{kWh})\left(300\frac{\text{day}}{\text{year}}\right) = 600\frac{\text{kWh}}{\text{year}}\end{aligned} \tag{7}$$

The total energy that PV system can save is 600kWh/year and based on this saving now we can calculate the total money that we can save annually [11].

$$\begin{aligned}\text{annually benefits} &= \left(\frac{\text{energy generation}}{\text{year}}\right)\left(\text{energy price}\frac{\text{AFN}}{\text{kWh}}\right)\\ &= \left(600\frac{\text{kWh}}{\text{year}}\right)\left(7\frac{\text{AFN}}{\text{kWh}}\right)\\ &= 4200\frac{\text{AFN}}{\text{year}}\end{aligned} \tag{8}$$

The operation and maintenance (O&M) cost of PV systems is 2.5% of investment cost. So, the O&M cost of the project is 100 AFN/year. Lifetime of the PV system is 25 years. The total benefit of the whole lifetime of the system is 110,000 AFN. In addition to the above benefits, the other advantages are as follows:

- Environmental benefits
- Sustainability of the grid
- Less outage.

As we know, the advantage of this project is not only for the consumer but also by applying this project there is a lot of benefits for DABS, so the government must have subsidies for applying this project for consumers because it increases the sustainability of the grid.

3.3 Payback Period

The payback time of investment in PV systems is calculated as follows [11]:

$$\begin{aligned}\text{Payback time} &= \frac{(\text{investment cost})}{(\text{yearly benefit cost}) - (\text{O\&M cost})}\\ &= \frac{18,500\,\text{AFN}}{4200\frac{\text{AFN}}{\text{year}} - 100\frac{\text{AFN}}{\text{year}}} \cong 4.5\text{ year}\end{aligned} \tag{9}$$

In addition, solar PV system is the best solution for the implementation of EE on lighting and light appliance. There are other methods of implementing EE in residential buildings in the lighting system. These methods are as follows:

- Installation of Light Emitting Diodes (LEDs) in place of incandescent lamps
- Using an automated device, such as a key tag system, to regulate the electric power in a room
- Behavioral change
- Improving building design to maximize natural light
- Policy option to help promote lighting efficiency.

3.4 Installation of Light Emitting Diodes (LEDs) in Place of Incandescent Lamps

Incandescent lamps are rarely used today. The impact of using these lamps is salient; however, it is used less than in the last decades in Kabul city. A way to decrease the impacts of incandescent lamps is replacing these lamps with LEDs. LEDs use a different, more advanced technology than incandescent light bulbs and come in a range of styles and sizes based on brand and purpose. LEDs use about 1/9 less energy than standard incandescent bulbs, give the same amount of light, and can last 6–10 times longer [12].

4 Heating System

As mentioned in previous chapters, buildings consume around 40% of the world's total energy consumption, and heating system is the largest part of energy consumption in residential buildings with about 30% to 40% and a corresponding significant amount of CO_2 emission [13]. EE in heating systems has become the key driver of sustainable development in many economies in the world. The availability and the use of energy in a building are pivotal to the building's functionality within the confines of its purpose.

Afghanistan imports 77% of its energy from the neighboring countries and residential buildings consume 71% of the total energy consumed [6]. Most the energy is consumed for heating in buildings because there is a significant heat loss in the winter season. Insufficient knowledge of the thermal performance of building structure is the main reason behind this situation. The EE in buildings goes through the Envelop, Geometry system, and the behavior of users.

The main sources of energy for the heating system in Kabul city are wood, coal, gas, oil, and electricity (Fig. 2). Consumption of fossil fuel is unfriendly with environment, and inefficient use of energy impacts the economy of society adversely.

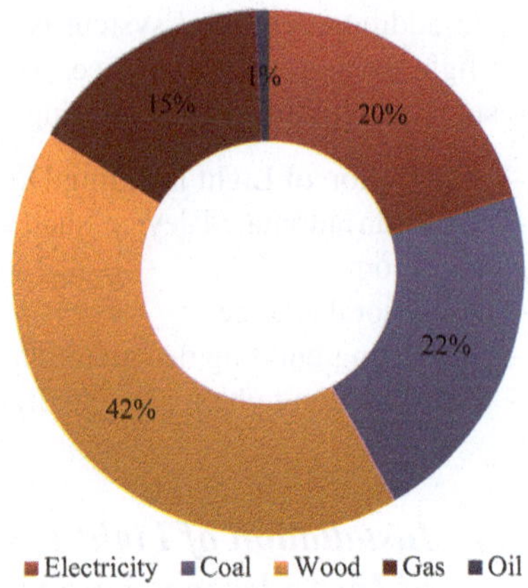

Fig. 2 Heating source in Kabul city [9]

The beneficial effects of energy efficiency on living conditions, on the reduction of fuel consumption, and on the environment can be linked to the Afghan specificities. The climate is semiarid, with a mountainous and continental climate, harsh winters, hot summers, and scarce rainfalls. A heating system is defined as any piece or combination of equipment that is used to raise the temperature in your home. This can be accomplished in several ways, using energy sources such as wood, coal, gas, oil, and electricity.

4.1 Types of Heating Systems

Regardless of the fuel source, heating systems can be divided into two basic categories.

- Space heating
- Central heating.

4.2 Heating and Efficiency

Energy efficiency does not mean that energy should not be used but that energy should be used in a manner that will minimize the amount of energy needed to provide services (decrease heat loss). There are a lot of EE methods for heating systems in the world, but all of these methods are not applicable in Afghanistan. As we mentioned before, some conventional EE methods are applicable in Kabul city because more than 20,000 buildings are unplanned, and more than 70% of buildings are unstandardized in the 11th district. Renewable Energy and Solidarity Group or Groupe Energies Renouvelables, Environnement et Solidarité (GERES) has studied 3000 households in different regions of Kabul city by the name of Live Comfortably,

Lower Heating Costs and Clean Indoor Air in 2015. In this study, some conventional and proper EE methods in heating systems are performed. Based on the GERES study [14], we accomplish these conventional methods for the 11th district. The technologies in this catalog provide an opportunity to improve the energy efficiency of already existing houses. It should be possible to save from 30 to 50% of fuel used for heating, to lower energy bills and reduce CO_2 output.

Three main proper indicators of this study for the 11th district of Kabul city are these conventional and easily handed EE methods for heating system:

- Double glazing;
- Inside roof insulation;
- Outside roof insulation;
- Passive heating.

4.3 Double Glazing

Windows can be one of your household's most beautiful structures. Windows offer views, daylighting ventilation, and heat from the sun in the winter. Unluckily, they can also account for 10–25% of your heating bill by allowing heat out [15]. Double-glazed windows have two sheets of glass fitted into a window frame. A passive gas like argon is filled between the two glasses to increase insulation. The double-glazed glass window is perfect for Afghanistan, especially for central provinces because these provinces have moderate summer and cold winter. The double window glasses not only provide relief from risky temperatures outside but also insulate your space from the noise outdoor. They are also tough to break through and therefore provide you with ample security against burglars.

4.4 Double Glazing Application in the 11th District Households

As we mentioned before, windows are the main factor of heat loss in houses, as their large surface is in direct contact with outdoor air and cold temperature. Overglazing your windows enables you to keep the heat inside and to significantly improve the comfort in your living room while reducing fuel consumption.

We select an average home here in the 11th district that the owner of the household wants to make double glaze where the average area of double glazing of this house is 10 m^2. The initial cost of related facilities is less than 13 USD per m^2 and more than 30% fuel saving. Table 3 shows the initial cost, saving money, and the payback period of implementation of double glazing in the average household (10 m^2) in the 11th district.

Table 3 Money saving, initial cost, and payback period of double glazing in the average home

Investment	Ranges
Saving per year	From 40 to 80 USD
Financial investment	Less than 150 USD
Duration of return on investment	Less than 2 years

Table 4 Money-saving, initial cost, and the payback period of double glazing in whole households in the 11th district

Investment	Ranges
Saving per year	From 1,520,000 to 304,000 USD
Financial investment	Less than 5,700,000 USD
Duration of return on investment	Less than 2 years

Table 4 shows the initial cost, saving money, and a payback period of implementation of double glazing, if we implement this study in all 38,000 households of the 11th district.

5 Fuel Saving and Its Environmental Benefits

In this section, we analyze different type of fossil fuel (coal, wood, and gas) and electricity that is used for heating a household in the 11th district, after the implementation of double glazing. Four different cases are analyzed: coal, wood, gas, and electricity. Because it is difficult to achieve how much families consume these fossil fuels together for the heating system.

5.1 *Coal*

This study reckons that an average family consumes one ton bituminous coal for heating. Figure 3 shows if a household used just coal for heating.

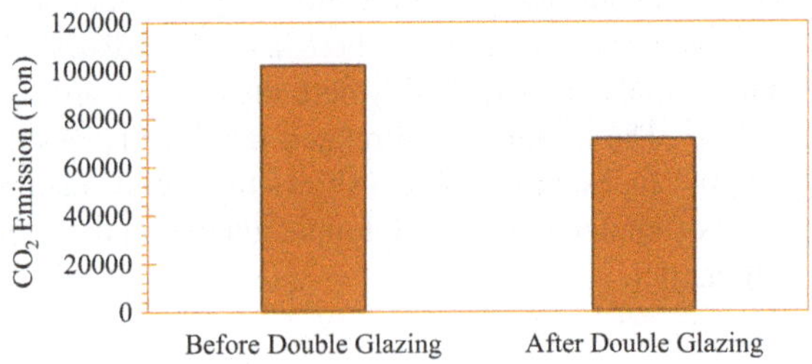

Fig. 3 CO_2 emission before and after the implementation of double glazing coal-based fuel

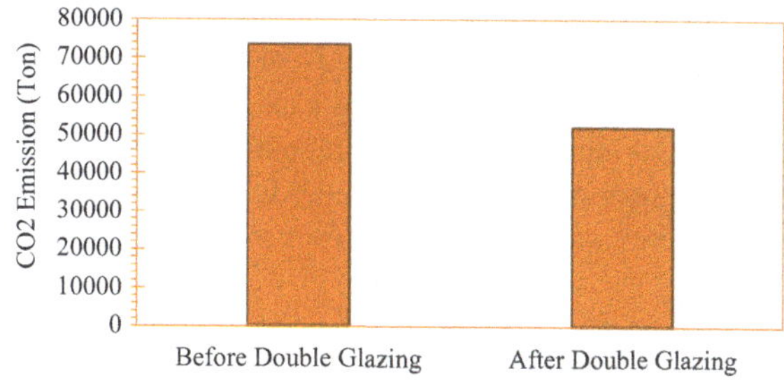

Fig. 4 CO_2 emission before and after implementation of double glazing wood-based fuel

5.2 Wood

Wood is one of the important fossil fuels used in Afghanistan. In this study we assume that an average family consumes 1120 kg wood for the heating system in the 11th district. Figure 4 shows if a household used just wood for heating.

5.3 Gas

Gas is also an important fossil fuel used in Kabul city, and gas has third position in usage after coal and wood (Fig. 5). In this study we assume that each family consumes 200 kg gas in the winter season.

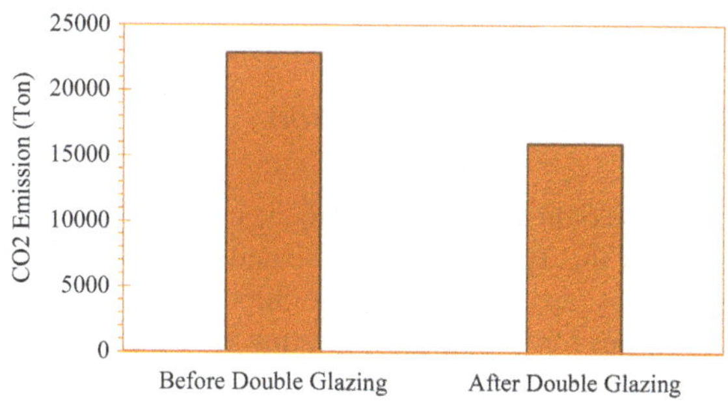

Fig. 5 CO_2 emission before and after implementation of double glazing gas-based fuel

Table 5 Money-saving, CO_2 emission decreasing, with each heating source coal, wood, and gas

Coal	Money-saving	114,000,000 AFN
	CO_2 decreasing	30,600 TON
Wood	Money-saving	182,418,000,000 AFN
	CO_2 decreasing	
Gas	Money-saving	114,000,000 AFN
	CO_2 decreasing	6,841.51 TON

Table 5 indicates money-saving options and environmental advantage of double glazing in all 38,000 households of the 11th district by the usage of fossil fuels generated from coal, wood, and gas.

Fuel saving and its environmental benefits in other methods are the same with double glazing because all methods save an equal amount of about 30% of fuel consumption, and this study prevented its repetition in other methods.

6 Insulation

When homeowners begin to think about the high price of energy, the first thing that comes to mind is insulation. Applying EE in homes, especially insulation protects and separates the indoor temperature from the outdoor temperature. Insulation is a vital component in any home that aims to be energy-efficient and serves to slow and reduce heat transfer [16]. Three things should be considered before insulation:

- First, fresh air must be provided to the occupants.
- Secondly, all uncontrolled air leaks in the thermal envelope should be sealed.
- Thirdly, you must prevent moisture accumulation.

There are many types of insulation in buildings. Because of existing unplanned, old, and conventional buildings in the 11th district, only two conventional roof insulations are applicable in the 11th district buildings.

- Inside roof insulation
- Outside roof insulation.

6.1 Inside Roof Insulation

This type of insulation is used to the interior part of a building, and the main material used for this type is polystyrene and glass wool. The use of this method can save energy by up to 30%, and the initial cost is around 10 USD per m^2. The main conditions for inside roof insulation application are the following:

- Traditional roof with wood or metal beams.
- The targeted ceiling must be at the top of the house.
- Roof in good condition, with no leakage.
- The targeted room should be the living room used in winter.

According to the GERES study, Table 6 shows the initial cost, payback period, and money saving of this method for 17 m^2 inside roof insulation for an average household.

Table 7 shows the initial cost, saving money, and the payback period of implementation of inside roof insulation, if we implement this study in all 38,000 households of the 11th district.

Table 6 Saving money, initial cost, and the payback period of implementing inside roof insulation in an average home

Investment	Ranges
Saving per year	From 40 to 80 USD
Financial investment	150 to 200 USD
Duration of return on investment	From 2 to 5 years

Table 7 Saving money, initial cost, and the payback period of implementing inside roof insulation in all households of the 11th district

Investment	Ranges
Saving per year	From 1,520,000 to 304,000 USD
Financial investment	5,700,000 to 7,600,000 USD
Duration of return on investment	From 2 to 5 years

6.2 *Outside Roof Insulation*

Insulating the outside of the basement works well with damp proofing and foundation drainage. Insulation can act as a drainage layer, keeping surface, and groundwater away from the foundation [14]. To maximize the insulation properties of the traditional roof, a mix of mud (30%) and straw (70%), with a specific thickness of 12 cm after compression, will significantly reduce the heat losses. In Kabul city, it is less than 10 USD per m^2 for roof renovation; for an average home, refer to Table 8.

Table 9 shows the initial cost, saving money, and the payback period of implementation of outside roof insulation, if we implement this study in all 38,000 households of the 11th district.

Table 8 Saving money, initial cost, and the payback period of implementing outside roof insulation (average home)

Investment	Ranges
Saving per year	From 40 to 80 USD
Financial investment	Less than 150 USD
Duration of return on investment	Less than 2 years

Table 9 Saving money, initial cost, and the payback period of implementing inside roof insulation in all households of the 11th district

Investment	Ranges
Saving per year	From 1,520,000 to 304,000 USD
Financial investment	About 5,700,000 USD
Duration of return on investment	Less than 2 years

6.3 Passive Heating

Passive solar heating is environmentally friendly. There is no electricity consumed with electric fans and no emissions produced by burning fossil fuels. In fact, when set up, passive systems consume no energy at all. Though these systems are most easily applicable when building a new home, some elements can be installed in many existing homes, especially when doing some major renovations.

Passive solar systems do not replace mechanical heating systems, but instead, act to reduce heating load requirements. Passive solar heating can reduce dependency on mechanical heating systems by 5–25% at almost no extra cost. More comprehensive passive heating systems can reduce dependency by 25–75% but have large initial investment requirements [17].

A passive solar heating system is a way for the building materials to collect, store, and distribute solar energy by natural convection, conduction, and radiation. The building itself acts as thermal mass to store the heat it collects during the day, which is then released during the night. Homes with high potential for solar electricity usually have good potential for passive solar heat. Passive heating systems in Kabul city decrease fuel consumption by at least 30% and 10% temperature increase inside the house. Here in Kabul city we will analyze a conventional passive heating system by the name of veranda passive heating. Because of the external features of the passive solar system, this system has necessary a little more financial cost. For an average household, refer to Table 10.

Table 11 shows the initial cost, saving money, and the payback period of implementation of implementing passive heating if we implement this study in all 38,000 households of the 11th district.

Table 10 Saving money, initial cost, and the payback period of implementing passive heating in an average home

Investment	Ranges
Saving per year	From 40 to 80 USD
Financial investment	From 250 to 400 USD
Duration of return on investment	From 2 to 5 years

Table 11 Saving money, initial cost, and the payback period of implementing passive heating in all households of the 11th district

Investment	Ranges
Saving per year	From 1,520,000 to 30,4000 USD
Financial investment	9,500,000 to 15,200,000 USD
Duration of return on investment	From 2 to 5 years

7 Result and Discussion

As we know, better energy is an important and vital issue in human life and without energy living on earth is impossible. Society development in Afghanistan brings increasing requirements on the quality of living. Nowadays, in times of lack of energy, we realize more and more than the price of fuel and energy has been continually increasing. We are asking ourselves where to find and how to realize saving solutions. Afghanistan imports 77% of its energy from the neighboring countries, and residential buildings consume about 71% of the total energy consumed. Lack of energy, CO_2 emission, and comfortable life are important reasons for applying EE measures in Afghanistan. Kabul city is among the sullied cities in the world. In Afghanistan, 23,000–26,000 dies of air pollution each year. Kabul city has the highest rate among all provinces. The 11th district of Kabul city is one of the populated districts in Kabul city, and applying EE measures helps in money saving and also has an important role in decreasing CO_2 emission in Kabul city.

EE applying in light appliances can save about 4200 AFN per year for an average household, whereas replacing incandescent lights with LED saves about 832.2 MWh/Year, including a salient role in decreasing carbon emission and reducing electricity shortage. EE in heating systems also has an important contribution in money saving and reduction of CO_2 emission; each of the abovementioned conventional EE measures in the heating system can save about 30% fuel. An average home can save up to 80$ annually, whereas it does not need much financial investment. In addition, CO_2 emission can reduce by up to 30%. For instance, if we can reduce consumption of coal 30%, we can reduce CO_2 emission from 102,000 to 71,400 tons; thus, it is very useful in reducing pollution in Kabul city and can rescue Kabul city from this calamity.

Briefly, we can say, comfort is an important and issue in life, and a way of making life comfortable is making buildings and households comfortable. An energy-efficient home will keep your family comfortable while saving money. Whether you take simple steps or make larger investments to make your home more efficient, you will see lower energy bills. Over time, those savings will typically pay for the cost of improvements and put money back in your pocket. Your home may also be more attractive to buyers when you sell.

Challenges always exist in all areas of life, and we must have solutions for these challenges. Here, in this study, these solutions are recommended for the related problems.

- Government subsidies
- Imposing of policies
- Advertisement for behavioral change.

7.1 Government Subsidies

As we know, it is difficult for a scant family to make an investment in EE sector. Although the initial investment of conventional EE and installation of the 400-W solar system is not too much, in Afghanistan, it is also difficult. For this reason, the government must presently have subsidies for applying these measures because by applying these measures, the government can increase the sustainability of the grid and reduce CO_2 emission.

7.2 Imposing Policies

For the effective implementation of energy efficiency projects, the development and formulation of an energy efficiency policy and the enactment of a legal and institutional framework are key actions that need to be taken. In Afghanistan, the Ministry of Energy and Water is the nodal institution responsible for the development and implementation of energy efficiency policies. The overall goal of the policy is to guide the Ministry of Energy and Water (MEW) on what it would like to achieve in terms of energy efficiency, how long it will take, and what it should do to achieve it. It is recommended that after the completion of the policy document, MEW should work on drafting a law to implement and enact this policy, which is the only means guaranteed to ensure its application. Some recommended policies are as follows:

- Incentive
- Appliance Standard
- Feed-In Tariff
- Education and outreach
- Energy prices should reflect real costs and give more incentives to consumers.

7.3 Advertisement for Behavioral Change of User

This advertisement must be in all sector of energy using, and EE must be applied in all division of residential buildings, including.

- Cooking
- Ventilation
- Cooling and freezing food
- Laundry boiling water
- Heating system
- Cooling system
- Using home electrical appliances
- Lighting system.

8 Conclusion

Energy is the base of improvement in society as, without daily energy, life is impossible. Afghanistan is a country that is faced with a lack of energy; there are several ways to decrease the problem of energy in Afghanistan. Applying EE is one of the ways of decreasing this problem. EE not only increases access to energy but also decreases environmental pollution. Application and practicing these proposed strategies and solutions can save up to 50% energy consumption with a short payback period of around three years. In addition to other economic benefits and environmental considerations, this research is proposing cost-effective and environmentally friendly solutions that could contribute towards the sustainability of the grid as well as ensuring access to the affordable, reliable, sustainable, and modern use of energy supply.

References

1. Nii, A.M., Emmanuel, A., Joshua, A.: developing a building energy efficiency assessment tool for office buildings in Ghana: Delphic consultation approach. Energy Procedia. **111**, 629–638 (2017). https://doi.org/10.1016/j.egypro.2017.03.225
2. Bataineh, K., Alrabee, A.: Improving the energy efficiency of the residential buildings in Jordan. Buildings. **8**(7), 85 (2018). https://doi.org/10.3390/buildings8070085
3. Mardookhy, M., Sawhney, R., Ji, S., Zhu, X., Zhou, W.: A study of energy efficiency in residential buildings in Knoxville, Tennessee. . J. Cleaner Production **85**, 241–249 (2014). https://doi.org/10.1016/j.jclepro.2013.09.025
4. Ershad, A.M., Sabory, N.R., Danish, M.S.S.: Energy efficiency guidebook. Deutsche Gesellschaft für internationale zusammenarbeit (GIZ) and Ministry of Energy and Water (MEW) - Afghanistan, Kabul, Afghanistan (2016)
5. Danish, M.S.S., Sabory, N.R., Ershad, A.M.: Renewable Energy Guidebook. Deutsche Gesellschaft für Internationale Zusammenarbeit (GIZ) and Ministry of Energy and Water (MEW) - Afghanistan, Kabul, Afghanistan (2016)
6. Temori, MO.: Energy efficiency guidebook for buildings/Mohammad Omar Temori ; design and layout Ahmad Sear Sharifi. Deutsche Gesellschaft für Internationale Zusammenarbeit (GIZ), Kabul, Afghanistan (2017)
7. Accelerating the Global Adoption of: Energy-Efficient Lighting. UN Environment, Nairobi, Kenya (17)AD
8. Zaidy, M.I.: Renewable energy efficient solar lighting system design for buildings in UAE. Int Res J Elec Comput Eng **2**, 13–17 (2016)
9. Waseel, A.M.: Central Heating by Seasonal Sensible Heat Storage of Solar Thermal Energy (2019)
10. Project profile on solar light systems: https://www.dcmsme.gov.in/reports/electronic/SOLARLIGHTSYSTEMS.pdf (2010)
11. Hagen, K.D.: Introduction to Renewable Energy for Engineers. Pearson, Hoboken, New Jersey (2015)
12. Efficiency, L.: PEW Center—Global Climate Change. Washington DC, USA (2011)
13. Olesen, B.W.: Heating and cooling systems for better energy efficiency. In: Proceedings of Clima 2007 WellBeing Indoors. pp. 45–60. he Federation of European Heating and Air-Conditioning Associations, Helsinki, Finland (2007)

14. Live comfortably: Lower heating costs, cleaner indoor air. Groupe Energies Renouvelables, Environnement et Solidarités (GERES), Kabul, Afghanistan (2018)
15. Lohia, S., Dixit, S.: Energy Conservation using Window Glazing in India. Int. J. Adv. Res. Electr. Electron. Instrum. Eng. **4**, 8645–8654 (2015). https://doi.org/10.15662/IJAREEIE.2015.0411016.
16. Insulation: https://www2.illinois.gov/sites/KeepWarm/Documents/insulation.pdf (2014)
17. Siliski, A., Foxworthy, R.: Everyday Environmental Stewardship, https://www.mipandl.org/ees.html, Last Accessed 24 Apr 2020

The manufacturer's authorised representative in the EU is Springer Nature Customer Service Centre GmbH, Europaplatz 3, 69115 Heidelberg, Germany. If you have any concerns regarding our products, please contact ProductSafety@springernature.com

Printed and bound by CPI Group (UK) Ltd, Croydon, CR0 4YY
30/09/2026
02225333-0005